Cause and Prevention of Clogging of Wells Abstracting Groundwater from Unconsolidated Aquifers

KWR Watercycle Research Institute Series

Cause and Prevention of Clogging of Wells Abstracting Groundwater from Unconsolidated Aquifers

C.G.E.M. (Kees) van Beek

Publishing
London • New York

Published by **IWA Publishing**
Alliance House
12 Caxton Street
London SW1H 0QS, UK
Telephone: +44 (0)20 7654 5500
Fax: +44 (0)20 7654 5555
Email: publications@iwap.co.uk
Web: www.iwapublishing.com

First published 2012

British Library Cataloguing in Publication Data
A CIP catalogue record for this book is available from the British Library

Library of Congress Cataloging-in-Publication Data
A catalog record for this book is available from the Library of Congress

ISBN: 9781780400242
ISBN: 1780400241

Contents

Preface

Groundwater has been abstracted with the help of drilled wells for over 100 years. It could be thought that this long history would result in traditional practical and theoretical advances. However nothing is less true: during the last years there has been considerable progress in well management, encompassing well design, well construction and well operation. This book summarizes the developments in these fields providing practical instructions based on sound theoretical insights. Consequently, Dutch water abstractors have changed their method of well construction and well operation, resulting in the near absence of well rehabilitation, the longer life of wells and increased abstraction reliability.

The author has more than 30 years experience in the field of groundwater abstraction. He started this book after his retirement from KWR Watercycle Research Institute (formerly Kiwa Water Research). The data used in this book were collected during his service in the Joint Research Programme of the water sector (BTO) by KWR Watercycle Research Institute and participating water companies and contract research for individual water companies and industries abstracting groundwater. This research benefitted much from discussions with colleagues and former colleagues from KWR and with colleagues from water utilities and industry. Besides these colleagues, the author is indebted to J.W. Kooiman as head of the Geohydrology Group and to prof. Dr. P.J. Stuyfzand who supervised this research in order to act as a Ph.D. thesis. At last, the assistance of ASPEX Corporation, Delmont, PA, USA, in providing the SEM/EDS data is gratefully acknowledged.

Summary

Groundwater serves many purposes: as raw material for the public and private drinking water supply, for the production of beer and beverages, and in agriculture for irrigation and watering cattle. Groundwater abstraction serves for stabilizing structures in civil engineering, like dams and tunnels, for keeping dry building excavations, and for reclamation of polluted groundwater. At the moment, groundwater is also increasingly used for energy storage (Aquifer Thermal Energy Storage, ATES).

Groundwater is usually abstracted with the help of wells. At present there are probably millions of drilled wells, either permanently or temporarily. In unconsolidated sediments these drilled wells consist of a well screen with on top a well casing. The annular space between the well screen and the well bore is filled up with gravel, the so-called gravel pack. In order to prevent short-circuit groundwater flow, the gravel pack is sealed on top with clay. For the same reason pierced clay deposits are also restored with clay. After construction, the well has to be developed in order to remove the drilling mud, which is plastered on the well bore and invaded a short distance into the aquifer. But remarkably, the criterion for a developed well consists of the absence of sand in the abstracted groundwater at a discharge rate usually twice the design capacity, and not of a successful removal of drilling mud.

After some time, drilled wells may become clogged. Clogging is characterized by a decrease in the specific capacity: the discharge rate (m^3/h) divided by the drawdown (m), where the drawdown equals the difference in water level in the well during operation and rest. In well clogging, two causes may be distinguished: clogging of the slots of the screen (screen slot or chemical clogging) and clogging of the well bore (well bore or mechanical clogging). Both types are easy to distinguish, as screen slot clogging is characterized by an increasing difference in water level during abstraction inside and outside the well, and well bore clogging by the absence of an increase of this difference.

Screen slot clogging

Screen slot clogging is caused by an accumulation of chemical precipitates, whether or not accompanied by biomass, in the screen slots. Most frequently, these precipitates consist of iron(III) hydroxides. These iron (III) hydroxide precipitates are not confined to the screen slots, but are also present upstream in the gravel pack, and downstream on the well casing and submersible pump and in the field lines. Precipitates will develop if the well abstracts two incompatible water qualities: in the case of iron(III) hydroxide precipitates shallow groundwater containing oxygen and deeper groundwater containing iron. In iron(III) hydroxide precipitation three oxidation processes may be active: homogeneous, heterogeneous and biological oxidation. In homogeneous oxidation, iron(II) and oxygen are mixed together and iron(III) hydroxides will precipitate. This process is slow under neutral conditions, and the rate increases with increasing pH and increasing iron(II) and oxygen concentration. As soon as some iron(III) hydroxide precipitates are present, heterogeneous oxidation will occur, characterized by adsorption of dissolved iron(II) by these precipitates and subsequent oxidation and hydrolysis of adsorbed iron(II). In this two stage process, the adsorption process is actually rate limiting. Under neutral conditions adsorption is fast, and the rate increases with increasing pH, iron(II) concentration and iron(III) hydroxide concentration, or better with increasing iron(III) hydroxide surface area. Everywhere where conditions are at least acceptable, bacteria adapted to those conditions will be present. Consequently, aquifers containing both oxygen and iron(II) will contain Iron Oxidizing Bacteria (IOB), which sooner or later will be attracted by the discharging well. As soon as IOB are present, biological oxidation will occur, characterized by uptake and adsorption of dissolved iron(II) on their cell walls and on their Extracellular Polymeric Substance (EPS) and subsequent oxidation and hydrolysis of iron(II). The physiological optimum conditions for IOB are characterized by slightly acid to neutral conditions and a low oxygen concentration. If both iron(III) hydroxide precipitates and IOB are present, there is a fierce competition between heterogeneous and biological adsorption of dissolved iron(II). In this reasoning the adsorption rate of dissolved iron(II) by on the one side iron(III) hydroxide surfaces and on the other side the uptake or adorption rate by biological surfaces is critical, and not so much the oxidation rate of adsorbed iron(II).

Initially, in a new well, only homogeneous iron(II) oxidation will occur, and this is a slow process. Possibly developed precipitates will be carried away with the abstracted groundwater. This situation changes radically during periods of rest: during these periods there is ample time for reaction, and moreover, flocs, if accumulated on the screen of the submersible pump, may twirl down, even deep down in the well. In addition, in the process of filling in the cone of depression around the well, water may flow out the well through the slots into the gravel pack. If present, iron(III) hydroxide flocs may then precipitate and accumulate in the slots and gravel pack. As soon as abstraction starts again, the iron (III) hydroxide precipitation process is accelerated by the occurrence of heterogeneous oxidation. IOB have excellent adhering properties. As soon as they have reached the well, they will stick there, in particular during periods of rest.

In this reasoning, screen slot clogging may be prevented by positioning the well screen over a depth interval with homogeneous chemical composition. If this is not possible, clogging may be retarded by continuous abstraction, whether or not with the help of a variable discharge pump, by taking advantage of the slow reaction kinetics. As a result of these slow reaction kinetics, precipitates will not precipitate in the screen slots, but in the submersible pump and field lines. This recommendation not only applies to the development of iron(III) hydroxide precipitates, but to all precipitates, with slow reaction kinetics.

Continuous abstraction as a preventive measure was already known by experience, but this reasoning gives some background for this phenomenon.

Well bore clogging

In well bore clogging, particles have been suspect for a long time, but it was not possible to ascertain their presence, let alone their role. Moreover, as the aquifer may be considered as a giant filter, their presence was not taken for granted. Once there was an opportunity to count particles in abstracted groundwater with a commercially available particle counter, their presence could be clearly demonstrated.

In these particle counters a laser beam is directed through a flow through cell, and the number of shadows, together with their size, is counted on the opposite side with the help of a sensor. In this way the particle concentration together with the corresponding particle size distribution, is obtained, here with a minimum diameter of 2 μm. Obviously, shadows are projected by all opaque materials, whether it be minerals, organic matter, bacteria or iron(III) hydroxide precipitates, but also by gas bubbles. In order to be sure that all counted particles originated from the aquifer, in this study particle countings have been limited to clean wells. Abstracted groundwater from these wells contains no oxygen, and the low sulfate concentration points towards the occurrence of, past or active, sulfate reduction.

Particle countings were executed on abstracted groundwater under normal operation conditions, obtained via the sample valve on the well head. Particle concentrations always showed a peak after switching on the well, which decreased quickly to more or less stationary values. These stationary concentrations range between 2 and 120 particles/ml, with 1 to 3% of the particles >15 μm. Particle concentration and particle size distribution were dependent on the abstraction rate: the higher the rate, the higher the concentration and the coarser the particles. Particle size distributions in water are usually described with the help of the Pareto distribution. However, compared to the Pareto distribution, in abstracted groundwater particles >7 μm were underrepresented.

Particle concentrations in abstracted groundwater are much lower than concentrations measured in groundwater obtained from monitor wells. Apparently, the particle supply around abstraction wells becomes exhausted. This exhaustion may be explained by the recurrent mobilization of particles each time the pump is started, of which the intensity decreases with increasing distance from the well. Filtration theory shows that under average groundwater flow conditions, mineral particles with an average diameter of circa 0.5 μm are most mobile. Smaller particles are less mobile due to diffusion, and larger particles due to sedimentation. This limited mobility of larger particles may also explain the under representation of particles >7 μm in abstracted groundwater.

Peak concentrations amounted to several thousand particles/ml and these concentrations appeared to originate from the well bore. Apparently, particles accumulated on the well bore during the preceding abstraction period, and are mobilized by the acceleration exerted by the water set in motion after the pump is switched on again. This behaviour was demonstrated by all wells, clogging and non clogging. Comparison of well operation mode and drawdown data learned that long continuous abstractions resulted in an increase in drawdown, which indicates the occurrence of well bore clogging. Apparently, a starting pump is able to mobilize a certain number of particles. If, on average, during preceding abstraction periods more particles are accumulated then switching on the submersible pump is able to mobilize, the well will start to clog. And, on the contrary, if on average less particles are accumulated than a starting pump is able to mobilize, there will be no well bore clogging. Consequently, well bore clogging may be prevented by intermittent abstraction. This prevention method was confirmed by results of field measurements.

Remarkably, the abstracted groundwater from all wells contains particles, but not all wells clog. Possibly, well operation of these wells is such that clogging does not occur. It is also possible that the abstracted groundwater contains less particles and/or the well bore is less susceptible to accumulation of particles, for instance as a result of a more complete well development, a larger well diameter or a coarser aquifer.

Thus, prevention methods for both types of well clogging are just the opposite: continuous operation for wells susceptible to screen slot clogging, and intermittent operation for wells susceptible to well bore clogging.

Particle mineralogy

Information on morphology and chemical composition of particles was obtained with the help of automated Scanning Electron Microscope – Energy Dispersive X-ray Spectroscopy (SEM-EDS). Particles were identified as minerals by comparing the measured chemical composition with the average chemical composition of minerals from handbooks. The greater part of the particles consisted of quartz and calcite with feldspars, glimmers and clay minerals and a small contribution of pyrite. Iron sulphide particles were clearly smaller than other particles. No iron hydroxide particles were present. Apparently, the particle mineral composition is governed by the mineral composition of the aquifer, as source of particles, and by the groundwater chemical composition, as controlling the chemical stability. It was not possible to identify in a direct way organic matter particles, but there were only a few particles containing no cation and/or silicon.

Compared to the mineral composition of the aquifer grains, calcite was overrepresented in particle composition. As the particle size distribution and the morphology of the calcite particles were not different from those of the quartz particles, this overrepresentation cannot be explained by morphological differences. This difference can also not be explained by dissolution of calcite aquifer grains or calcite coatings as groundwater was saturated or over saturated with respect to calcite. Oversaturation may lead to precipitates, but calcite particles were larger than those of pyrite, which are definitely generated by oversaturation. Moreover, not all ground waters were oversaturated with respect to calcite. This leaves crushing of calcite grains under the weight of the overlying soil layers, and by the vibrations exerted by starting pumps, as the most likely cause. This also explains the overrepresentation with respect to quartz: quartz is much harder than calcite and consequently much harder to crush.

Most surprising was the presence, up to 25%, of particles consisting of sulfides of copper, lead and zinc, and particles consisting of iron whether or not containing chromium and nickel and of aluminum. These particles are considered as corrosion products from the well and from the sample valve on the well head, and as abrasion products from the submersible pump.

Well incrustation and well design criteria

At last, well design criteria in relation to incrustation were evaluated. These criteria are not correct and internally conflicting: in order to prevent incrustation of wells they aim to prevent turbulent flow. But aside from better mixing, turbulent flow has nothing to do with chemical processes: as long as no solubility products are exceeded and fast reaction kinetics are involved, no precipitates will develop. On the other side these criteria permit higher flow velocities in coarser aquifers. But the coarser the aquifer, the lower the flow velocity where laminar flow changes into turbulent flow.

Probably higher design velocities may be applied, allowing at least incipient turbulent flow. For the construction of new wells, it is advised to make more use of experiences from existing wells or from related well fields.

Chapter 1
General Introduction

C.G.E.M. van Beek

KWR Watercycle Research Institute, Groningenhaven 7, 3433 PE Nieuwegein, the Netherlands

1.1 BACKGROUND

Groundwater is an important resource for the public and private drinking water supply, for the production of beer and beverages, for paper making and in laundries, and for agriculture for irrigation and watering cattle. Groundwater is also important as an energy carrier for air conditioning and recently also for Aquifer Thermal Energy Storage (ATES) and as a cooling agent in industry. In civil engineering groundwater abstraction is important for pressure relief of dams and constructions, for prevention of landslides in mountainous areas (Buchacher et al. 2009), for draining construction sites and for reclamation of polluted groundwater. Where unconsolidated sediments are involved, groundwater is generally abstracted with the help of drilled wells. These wells are actually long pipes (=well casing) with a perforated bottom end (=well screen). The annular space between screen and aquifer is filled up with gravel and along the casing with clay and excavated material. Groundwater abstraction wells may clog by several causes. This clogging will become prominent in particular under unfavorable conditions, such as during dry periods with peaking drinking water demands and corresponding low groundwater levels, or during flood periods with dams stemming high surface water levels. Consequently, properly functioning wells are of vital importance.

As a large part of the world population is living in sedimentary areas, the greater number of wells abstracts groundwater from unconsolidated sediments. And because there are millions of drilled wells, even a small improvement in their performance results in greater operational reliability and in great financial advantages. Moreover, less frequent rehabilitation with chemicals and less frequent replacement by new wells with application of drilling fluids and traffic to and from drilling sites also results in less environmental damage.

Clogging is the greater cause of malfunctioning of wells. Clogged wells may be rehabilitated a number of times, after which they have to be replaced.

1.2 CLOGGING OF WELLS

Clogging of wells is as old as the drilling of screened wells. There is an extensive literature on well rehabilitation (Houben and Treskatis 2007), where the first papers are more than 100 years old (van Beek and Brandes 1977). Clogging is defined as a decrease over time of the specific discharge rate, where the specific discharge rate equals the discharge rate over drawdown, and where drawdown is defined as the difference in water level during abstraction and during rest. In well clogging two types may be distinguished: screen slot (or chemical) and well bore (or mechanical) clogging. Table 1.1 summarizes characteristics of both types.

Table 1.1 Characteristics of screen slot (or chemical) and well bore (or mechanical) clogging

	Screen slot clogging	**Well bore clogging**
Characteristic difference	Entrance resistance	No entrance resistance
Cause	Accumulation of chemical precipitates (and biomass) in the screen slots	Accumulation of particles on the well bore
Phenomenological characteristic	Well, pump and discharge lines completely covered with chemical precipitates, whether or together with biomass (slimes)	Well, pump and discharge lines completely clean
Chemical characteristic	Abstracted groundwater contains iron (manganese, methane) as well as oxygen (nitrate)	Abstracted groundwater shows occurrence of sulfate reduction?
Aquifer characteristic (in the Netherlands)	Abstraction from phreatic aquifers	Abstraction from (semi-) confined aquifers
Geology	Simultaneous abstraction from brown and grey colored sediments	Abstraction from grey colored sediments
Rehabilitation	Easy to rehabilitate as the clogging is easy accessible, and as long as they are fresh	Hard to rehabilitate, as the clogging is at some distance from the well and difficult to reach

Both types of well clogging are conveniently distinguished with the the help of the entrance resistance: screen slot clogging being characterized by the presence and well bore clogging by the absence of an entrance resistance, where entrance resistance is defined as the difference in water level in the well and in the monitor well in the gravel pack during abstraction. The other items may be considered as additional characteristics, supporting the diagnosis of the cause of well clogging.

Clogging rates are usually expressed as a decrease in specific capacity per year or as an increase in drawdown, at constant discharge rate, per year. However, these clogging rates are mutually hard to compare as there may be large differences in well operation, for instance by the number of operation hours per year and the length of the individual abstraction periods, and well construction, for instance by screen diameter and screen length, and pump capacity.

1.2.1 Screen slot clogging

Screen slot clogging is caused by an accumulation of chemical precipitates, whether or not accompanied by biomass. These precipitates consist usually of iron hydroxides with all kinds of admixtures, like phosphate, manganese and trace elements (Houben 2003). In the biomass iron oxidizing bacteria like *Gallionella ferruginea* and *Lepthotrix ochracea* are easy to recognize. Upon aging, under thick biomass layers, anaerobic conditions may develop, giving rise to all kinds of secondary reactions.

Research into screen slot clogging was aroused by a paper by Houben (2006) where the occurrence of iron hydroxide precipitates in the gravel pack was described. This occurrence was confirmed by nice pictures of excavated wells (Houben and Treskatis 2007) showing the gravel pack completely cemented by iron hydroxides, but not the adjacent aquifer. It is not clear whether this gravel pack is cemented completely over its whole length or only partly. The presence of iron hydroxide precipitates in the gravel pack is intriguing, and in order to explain this presence groundwater chemistry, well hydraulics, well operation and well construction were considered in detail. This approach led to a well operation method which prevents, or at least mitigates considerably, screen slot clogging.

1.2.2 Well bore clogging

In well bore clogging groundwater flow towards the well during abstraction is hampered at some distance from the well at the well bore. Apparently, due to the increase in groundwater flow velocity one or more processes in the aquifer, that interfere with the well bore, proceed with greater intensity. Based on reaction products in spent water after rehabilitation with sodium hypochlorite this clogging was ascribed to an accumulation of iron sulfides and biomass of sulfate reducing bacteria, with the possibility of accumulation of particles (van Beek and Kooper 1980). Later, active sulfate reduction was confirmed by the presence of sulfate reducing bacteria (van Beek and van der Kooij 1982). However, afterwards it appeared that iron sulfides were natural constituents of these aquifers (van Beek et al. 1989; Kölle et al. 1983), and that this type of clogging also occurred in aquifers without active sulfate reduction.

At that time, convenient ways to get an impression of the presence of particles were not available. It was only a few years later that particle counters became commercially available. And indeed exploratory countings showed the presence of particles in (abstracted) groundwater. In order not to be confused by iron hydroxide or biomass particles these countings were limited to clean wells, abstracting anoxic (sulfate reducing) groundwater from (semi-)confined aquifers. As the lower particle diameter limit of the counter was equal to 2 μm, it were indeed particles, see Figure 1.1. Consequently, in this book "particles" will be discussed, but groundwater will also contain colloids. Units composing the aquifer matrix are called "grains".

In order to gain information on the presence of particles in abstracted groundwater, particle concentrations in many clogging and non clogging wells were counted.

Combination of information on particle concentrations with results of drawdown measurements and logs of well operation led to a well operation method which prevents well bore clogging. This method was presented at the international IWA congress in Beijing in 2006 and won the international innovation award for Europe.

1.3 OUTLINE OF THE BOOK

Until recently it was assumed that in screen slot (or chemical) clogging the occurrence of iron(III) hydroxide precipitates was limited to the screen slots and further downstream. But excavated wells showed that these iron(III) hydroxide precipitates were also present upstream over the whole width of the gravel pack, and even extending some distance in the aquifer (Houben and Weihe 2010). Apparently, a combination of hydraulic and chemical processes lead to this occurrence. Chapter 2 presents the physical chemistry of iron(III) hydroxide precipitation, relevant for iron(II) removal in drinking water treatment from groundwater, in natural systems like seeps and wetlands, and in treatment of Acid Mine Drainage (AMD).

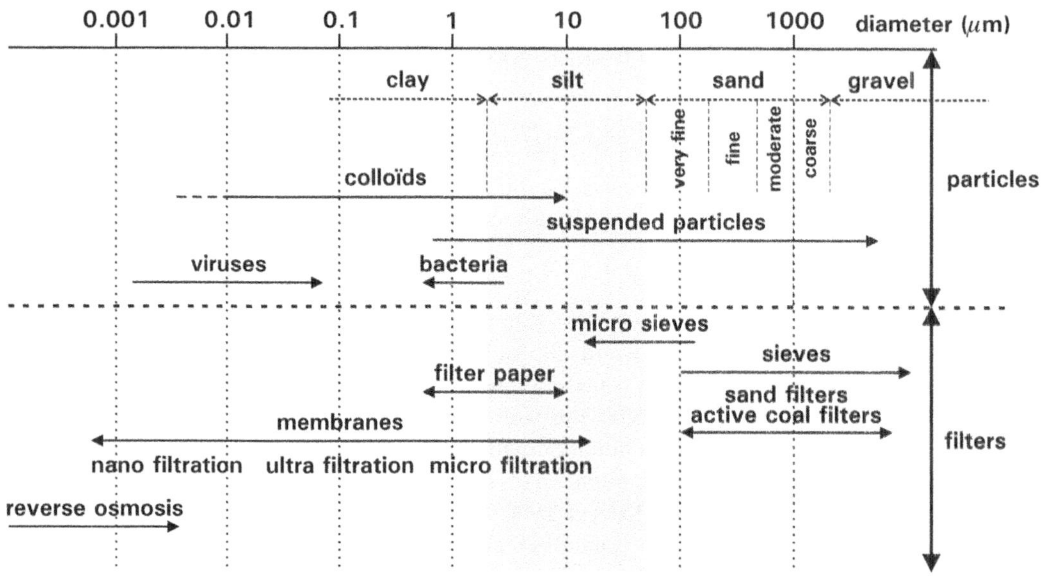

Figure 1.1 Summary of particle and filter diameter, modified after a.o. Stumm (1977). In this study particles are limited to the grey zone, bounded to the lower side by the minimum particle size measured by the particle counter, and to the upper side by the diameter of the pore throats of the aquifer

Particle concentrations in abstracted groundwater from many wells were measured. In order to see whether the presence of particles was relevant in clogging, these concentrations were measured in clogging and non clogging wells. The results of these countings are evaluated in Chapter 3. Moreover, particle concentration measurements also provide particle size distributions. These distributions are also evaluated and compared with results from surface water and from groundwater obtained from monitor wells.

In order to understand the behavior of particles in groundwater in aquifers, information about the mineralogy and the morphology of these particles is necessary. Chemical and morphological characteristics were obtained with the help of Scanning Electron Microscopy – Energy Dispersive X-ray Spectroscopy (SEM-EDS). The results of the chemical analyses were evaluated and particles were assigned to minerals by comparison of the experimental element ratios with standard ratios. These results are presented in Chapter 4.

By comparing the results of simultaneous measurements of particle concentration, well operation and drawdown over time, the cause of well bore clogging emerged, see Chapter 5. With the help of this cause, it was also possible to explain the phenomenon that some well(-field)s were clogging and others not, and to devise a method for prevention.

This prevention method has been applied now for several years on well field Tull en 't Waal (water utility Vitens) and the experiences were evaluated, see Chapter 6. As over this period only one well had to be rehabilitated, and before that time each well every 2 to 4 years, this prevention method proved effective.

As already mentioned, there are two types of well clogging. These two types ask for very different approaches with respect to prevention (and rehabilitation). Chapter 7 presents, for the oprator, a practical summary on identification, cause and prevention of screen slot and of well bore clogging.

In handbooks on well drilling and well maintenance time and again the same design criteria for wells are mentioned. Application of these design criteria would prevent well clogging by incrustation. Chapter 8 presents a detailed evaluation of these criteria. Moreover, the processes leading to screen slot clogging are described in detail, as are the processes leading to iron(III)hydroxide precipitation in the gravel pack and in the aquifer.

Chapter 9 presents a synthesis of the highlights of this study and connects some of the results of the various chapters.

Chapter 2

Review of homogeneous, heterogeneous, and biological oxidation of ferrous iron in drinking water treatment*

C.G.E.M. van Beek[1], T. Hiemstra[2], J.L.C.M. van de Vossenberg[1], P.S. Hofs[1], M.M. Nederlof[1], J.A.M. van Paassen[3] and G.K. Reijnen[4]

[1]*KWR Watercycle Research Institute, P.O. Box 1072, 3430 BB Nieuwegein, the Netherlands, Tel: +31 30 60 69 766, Fax: +31 30 60 61 165, E-mail: Kees.van.Beek@kwrwater.nl*

[2]*Wageningen University, Dept of Soil Quality, P.O. Box 47, 6700 AA Wageningen, the Netherlands*

[3]*Water utility Vitens, P.O. Box 1090, 8200 BB Lelystad, the Netherlands*

[4]*Water utility WML, P.O. Box 1060, 6201 BB Maastricht, the Netherlands*

ABSTRACT

Several processes are able to precipitate iron(II) as iron(III) hydroxides: homogeneous (=flocculent), heterogeneous (=adsorptive, auto-catalytic or contact) and biological oxidation. Here is evaluated under which circumstances each of these processes is dominant. With the help of physical chemistry it is demonstrated that in the presence of iron(III) hydroxide precipitates, heterogeneous oxidation is the dominant process, while homogeneous oxidation is negligible. As soon as Iron Oxidizing Bacteria (IOB) are present, biological oxidation may contribute substantially under slightly acidic conditions.

In literature the term "biological iron oxidation", called here "biological iron removal", in fact refers to a technological method, characterized by high filtration rates and limited oxygen supply, rather than to the process of biological iron oxidation by iron oxidizing bacteria (IOB). The contribution of biological iron oxidation to iron removal in biological iron removal plants is variable and may even be absent. In order to obtain an efficient removal of iron from groundwater in drinking water treatment, it is advised to optimize these plants for heterogeneous oxidation.

*Resubmitted after revision as: Kees van Beek, Tjisse Hiemstra, Jack van de Vossenberg, Bas Hofs, Maarten Nederlof, Jacques van Paassen and Gert Reijnen: Review of homogeneous, heterogeneous, and biological oxidation of ferrous iron in drinking water treatment.

2.1 INTRODUCTION

Removal of iron is one of the first steps in the treatment of groundwater to drinking water. This is usually achieved by oxidation of iron(II) to iron(III), followed by the formation of iron(III) hydroxides. On iron removal, recently a number of papers appeared, for instance El Azhar et al. (2008), de Ridder et al. (2008), DVGW (2005), Katsoyiannis and Zouboulis (2004), Pacini et al. (2005), Park and Dempsey (2005), Sharma et al. (2005, 2001), Groth and Czekalla (2004), Sharma (2001), Søgaard et al. (2001, 2000), Sommerfeld (1999) and Bruins et al. (1998). Oxidation and precipitation of iron are also active in screen slot (chemical) clogging (Stuyfzand, 2007; Houben and Treskatis, 2007).

Removal of iron from groundwater is not only important for the production of drinking water, but also for some industrial uses and for irrigation in glass houses in horticulture. A good understanding of iron removal has thus great economic value. Moreover, a literature survey learns that after the basic study of Sommerfeld (1999) no more recent reviews have been published in this field.

The present review starts with describing the solubility of iron(II) and iron(III). Kinetics play an essential role in the pathway of Fe(II) oxidation and precipitation as Fe(III) hydroxides after aeration. These will be described for homogeneous (=flocculent) and heterogeneous (=adsorptive, auto-catalytic, or contact-) oxidation. In addition Fe(II) oxidation by Iron Oxidizing Bacteria (IOB) will be depicted. Next, biological iron removal, in literature also called biological oxidation, will be re-evaluated in terms of heterogeneous and biological oxidation. Based on these results, recommendations will be presented for iron removal in drinking water treatment from groundwater.

This review does not include oxidation of iron(II) under acid conditions ($pH < 4$) by *Thiobacillus ferrooxidans*, as well as the oxidation by iron oxidizing photoautotrophic bacteria. These processes are not relevant for drinking water treatment.

2.2 IRON HYDROXIDES AND SOLUBILITY OF IRON

2.2.1 Mineral stability

Iron(III) hydroxides occur in series of minerals that differ in solubility and surface area. Figure 2.1 presents a schematic overview. The lines show various pathways of formation and transformation. Iron(III) hydroxides show variations in surface properties like charge density and pH_{ZPC} (Stumm and Morgan, 1981) that may affect the adsorption and surface oxidation of Fe(II) (Hiemstra and Van Riemsdijk 2007).

Usually, fast oxidation will produce Ferrihydrite or Hydrous Ferric Oxide (HFO). Upon aging, more stable compounds may form (Figure 2.1). For example, HFO rapidly converts into goethite under anoxic conditions in the presence of dissolved iron(II) (a.o. Park and Dempsey, 2005).

Fresh iron(III) hydroxide precipitates in clogging wells consist mainly of HFO, and older precipitates of goethite. Sometimes, Lepidocrocite occurs as a minor constituent, while the presence of hematite has never been reported (Houben and Treskatis, 2007). Iron(III) hydroxides, precipitated during iron removal from groundwater in drinking water treatment, consist of HFO (Carlson and Schwertmann, 1987), and in Subsoil Iron Removal (SIR) predominantly of goethite with some HFO (Mettler et al., 2001).

2.2.2 Solubility

The solubility of iron(II) and of iron(III) depends on many variables. For instance, for the pH range 3–13, the solubility of iron(III) in equilibrium with HFO is <1 μmol/L (=0.056 mg/L) (Stumm and Morgan, 1996).

In the presence of oxygen, the solubility of iron(II) in equilibrium with HFO is negligible. For instance, if the oxygen concentration equals 1 μmol/L (=0.032 mg/L), the iron(II) concentration will be >1 μmol/L only if pH < 1. In the absence of oxygen, the solubility of iron(II) is governed by $Fe(OH)_2(s)$ if pH > 8.85, and by $FeCO_3$ (siderite) if pH < 8.85 (Stumm and Morgan, 1996), and can be very high: >1 mol/L if pH < 6.

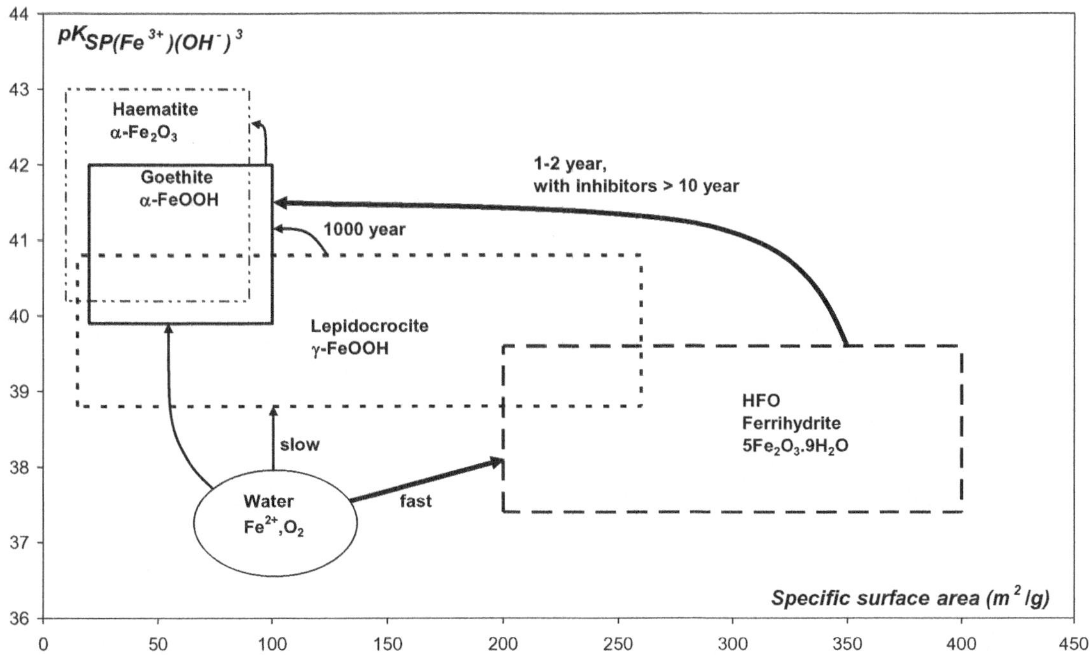

Figure 2.1 Relations between various iron(III) hydroxides (modified after Stuyfzand (2007), based on data from Houben and Treskatis (2007) and Appelo and Postma (2006). Arrows without number denote unknown rates

2.3 PRECIPITATION OF IRON(III) HYDROXIDES

As mentioned above, in the oxidation of iron(II) to iron(III) and further precipitation of iron(III) hydroxides, three processes may be distinguished (de Ridder et al., 2008): homogeneous (=flocculent), heterogeneous (=adsorptive, autocatalytic or contact) and biological oxidation.

2.3.1 Homogeneous oxidation

In homogeneous oxidation, iron(III) hydroxides are precipitated directly from dissolved iron(II) and oxygen:

$$\text{Oxidation:} \quad Fe^{2+} + \tfrac{1}{4}O_2 + H^+ \longrightarrow Fe^{3+} + \tfrac{1}{2}H_2O$$

$$\text{Hydrolysis:} \quad Fe^{3+} + 3H_2O \longrightarrow Fe(OH)_3\downarrow + 3H^+$$

$$\text{Sum:} \quad Fe^{2+} + \tfrac{1}{4}O_2 + 2\tfrac{1}{2}H_2O \longrightarrow Fe(OH)_3\downarrow + 2H^+ \tag{2.1}$$

Due to the presence of Fe(II) ion pairs besides Fe(II) ions, homogeneous oxidation is somewhat more complicated than represented here (Morgan and Lahav, 2007; Lerk, 1965), but this does not lead to other conclusions in our evaluation.

The rate equation for homogeneous oxidation may be represented as (Tamura et al., 1980; Sung and Morgan, 1980):

$$\frac{d}{dt}[Fe^{2+}]_{homo} = -k_1 \frac{[Fe^{2+}][O_2]}{[H^+]^2} \tag{2.2}$$

From rate equation (2.2), at constant oxygen concentration and pH, the half-life time may be calculated as (Appelo and Postma, 2006; Applin and Zhao, 1989):

$$t_{1/2} = \frac{\ln 2[H^+]^2}{k_1[O_2]} \tag{2.3}$$

The value for the rate constant k_1 at 25°C equals $1.08 * 10^{-13}$ mol L^{-1} s^{-1}. This value is very much dependent on the temperature. At 10°C, which is the average temperature of groundwater in The Netherlands, k_1 is exactly a factor 10 lower, $1.08 * 10^{-14}$ mol L^{-1} s^{-1} (Stuyfzand, 2007), which corresponds with a 10 times longer half-life time.

Figure 2.2 presents the half-life as function of oxygen concentration for various values of pH: the higher the oxygen concentration and the higher the pH, the smaller the half life time. This relationship is only approximate; the real relationship depends not only on temperature, but also on the presence of interfering compounds, like organic matter, phosphate, silicate, etc.

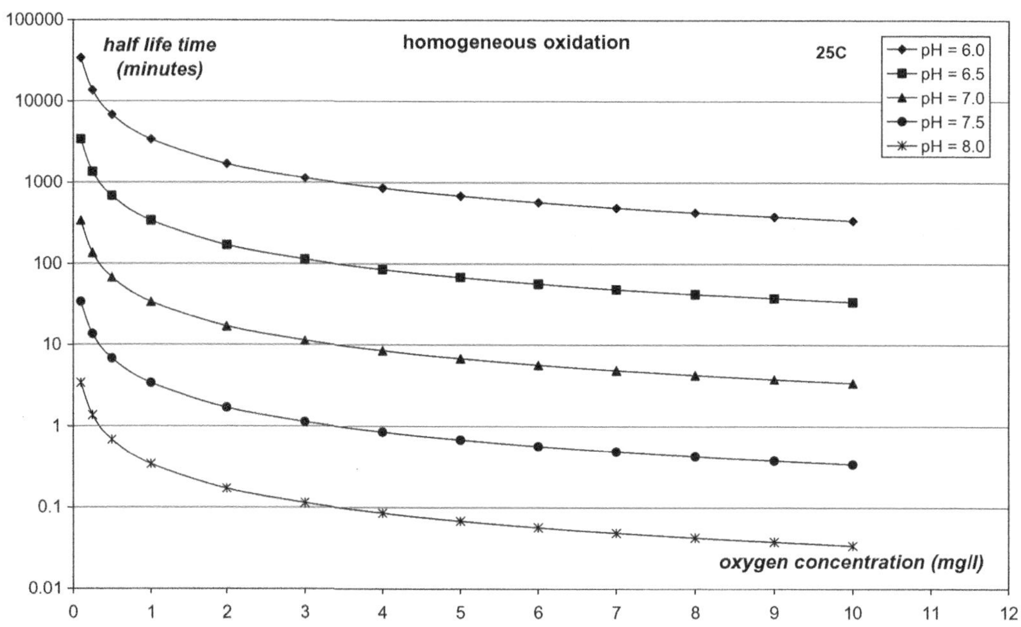

Figure 2.2 Half life time for the homogeneous oxidation of iron(II) as function of oxygen concentration for various values of pH at 25°C

2.3.2 Heterogeneous oxidation

In heterogeneous iron oxidation, two subsequent reaction steps can be distinguished. In the first step Fe(II) is adsorbed. The surface complex formed may transfer an electron to the mineral bulk (Hiemstra and van Riemsdijk, 2007), leading to an adsorbed Fe(III) surface complex that is unstable and will hydrolyze. This process can be described with variable charge models (Hiemstra and van Riemsdijk, 1996, 2007). Modeling shows that in the pH range of interest about one proton per adsorbed Fe^{2+} is released. In a simplified picture, one may use $S\text{–}OH^0$ sites to define thermodynamically the overall reaction. However, note that the actual mechanism is different (Hiemstra et al., 1996). With this choice of $S\text{–}OH^0$, the overall reaction of adsorption by the surface groups with proton release can be formulated as done in Eq. (2.4a).

$$\text{Fe(II)adsorption:}\quad S{-}OH^0 + Fe^{2+} \longrightarrow S{-}OFe(II)^+ + H^+ \tag{2.4a}$$

$$\text{Oxidation:}\quad S{-}OFe(II)^+ + \tfrac{1}{4}O_2 + 1\tfrac{1}{2}H_2O \longrightarrow S{-}OFe(III)(OH)_2^0 + H^+ \tag{2.4b}$$

$$\text{Overall:}\quad S{-}OH^0 + Fe^{2+} + \tfrac{1}{4}O_2 + 1\tfrac{1}{2}H_2O \longrightarrow S{-}OFe(III)(OH)_2^0 + 2H^+ \tag{2.4c}$$

In the second step of the heterogeneous iron oxidation, the hydrolyzed $S\text{–}OFe(II)^+$ surface complex will oxidize, releasing an additional proton. This process is formulated with Eq. (2.4b). The overall reaction shows that removal of each Fe^{2+} leads to the release of two protons and to the axidation and hydrolysis of adsorbed Fe(II). Continuous repetition of this last process will lead to crystal growth.

The rate equation is first order with respect to proton release. Considering the reactions of Eq. (2.4), the adsorption as well as the oxidation step may explain the rate law. Usually, ion adsorption is fast, as is the oxidation of adsorbed metals (Wehrli, 1990), where the adsorption rate also depends on pH and anion type (Jeon et al., 2003, 2001).

Heterogeneous oxidation may proceed along two pathways, i.e. in the continuous presence of oxygen by simultaneous adsorption and oxidation, as in the classical iron removal process, or in the discontinuous presence of oxygen by adsorption and oxidation separated in time, as in subsoil iron removal (SIR). Adsorption of iron(II) in SIR is confirmed by the presence of different iron(II) isotopic ratios in abstracted groundwater (Teutsch et al., 2005; Mikutta et al., 2009). Both mechanisms will be discussed next.

Simultaneous adsorption and oxidation

The rate equation for heterogeneous oxidation may be represented as (Tamura et al., 1980; Sung and Morgan, 1980):

$$\frac{d}{dt}[Fe^{2+}]_{hetero} = -k_2\frac{[S{-}OH^0][Fe^{2+}][O_2]}{[H^+]} \tag{2.5}$$

In this relation ($S\text{–}OH^0$) represents the concentration of iron(III) hydroxide, in mol/L. In the experiments of Tamura et al. (1976), pH ranged between 6.2 and 6.6 and iron(III) hydroxide concentrations ($S\text{–}OH^0$) were up to 30 mg/L. In the experiments of Sung and Morgan (1980), pH was 7.2 and iron(III) hydroxide concentrations were up to 3 mg/L. The value for k_2 depends on the structure of the iron(III) hydroxide, and ranges a factor 10 between the most and the least stable phase (Tamura et al., 1980): k_2 (HFO) $\gg$ k_2 (Goethite, Lepidocrocite), where the value for HFO equals $1.46 * 10^{-3}$ L mol^{-1} s^{-1} at 25°C. Mettler et al. (2001) and Mettler (2002) arrived at about the same rate sequence: HFO $\gg$ Goethite $>$ Lepidocrocite. Obviously, not the weight concentration of iron(III) hydroxides, in mol/L, is relevant but variables that refer to the mineral phase, like surface area, in m^2/L, or adsorption capacity, in meq/L.

This is confirmed by Tüfekci and Sarikaya (1996), who reported lower values for k_2 at higher iron(III) hydroxide concentrations.

In the same way as for homogeneous oxidation, the half life time for heterogeneous oxidation may be calculated as:

$$t_{1/2} = \frac{\ln 2}{k_2} \frac{[H^+]}{[S{-}OH^0][O_2]} \approx \frac{\ln 2}{k_2} \frac{[H^+]}{[S^*][O_2]} \tag{2.6}$$

In this relation $[S{-}OH^0]$ can be substituted by $[S^*]$. This is allowed as long as the amount of adsorbed iron (II) is small with respect to the adsorption capacity: $[S^*] = [S{-}OH^0] + [S{-}OFe(II)^+] \approx [S{-}OH^0]$. From Eq. (2.6) it is apparent that the half life time is equally dependent on proton, HFO and oxygen concentration. Figure 2.3 shows the half life time for heterogeneous oxidation of iron(II) as function of oxygen concentration for various values of pH and two iron(III) hydroxide concentrations.

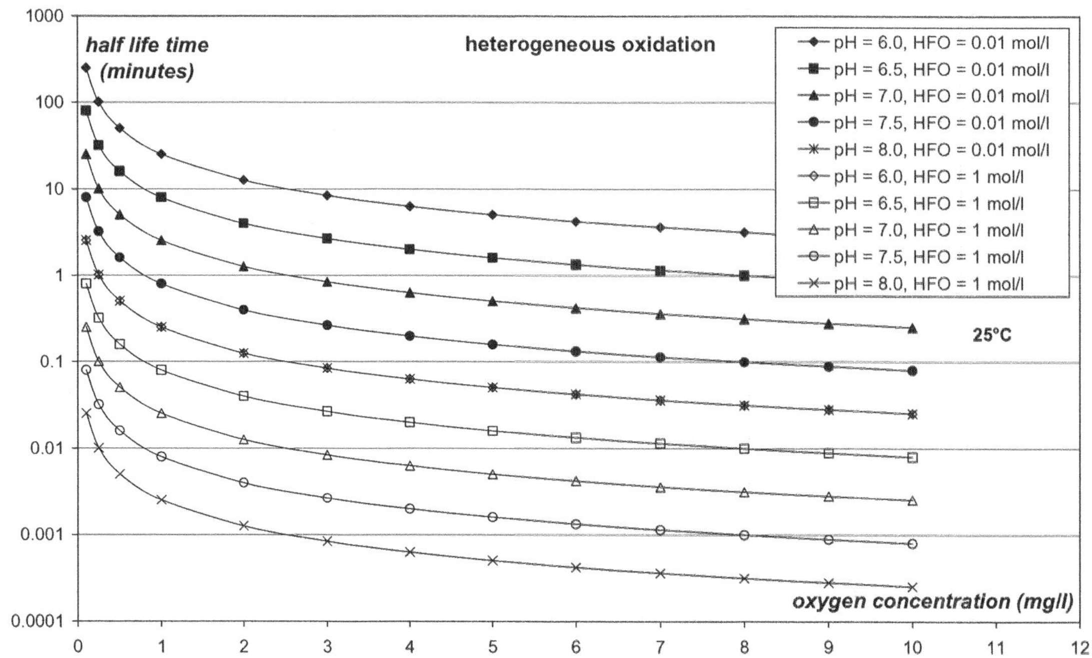

Figure 2.3 Half life time for the heterogeneous oxidation of iron(II) as function of oxygen concentration for various values of pH and two iron(III) hydroxide concentrations at 25°C

Adsorption and oxidation separated in time

Jeon et al. (2003) distinguish between adsorption and fixation of iron(II) by iron(III) hydroxide surfaces. This distinction is ascribed (Dixit and Hering, 2006) to the conversion of initial iron(III) hydroxide into mixed iron(II)/iron(III) oxide or to the transfer of electrons of adsorbed iron(II) to the bulk iron(III) hydroxides (Wehrli et al., 1989; Hiemstra and van Riemsdijk, 2007). Possibly this phenomenon also explains the very preferential adsorption of iron(II) by iron(III) hydroxides, as observed in SIR (van Beek, 1985, 1983, 1980).

Thermodynamically, the primary Fe(II) adsorption reaction (Eq. 2.4a) can be considered as Fe(II) adsorption with non stoichiometric release of one H^+ per Fe^{2+} for which in a non-electrostatic model approach a corresponding equilibrium can be defined as:

$$K_{Fe/H} = \frac{[S{-}OFe(II)^+][H^+]}{[S{-}OH^0][Fe^{2+}]} \tag{2.7}$$

in which the sites are expressed in mol HFO/L. The iron(II)-proton exchange constant on iron(III) hydroxide surfaces equals (Tamura et al., 1976): $K_{\mathrm{Fe/H}}=1.41 * 10^{-5}$. This value is identical to the value reported by Choi et al. (2001) for iron(II)–proton exchange on Lepidocrocite surfaces, and comparable to the value reported by Wolthoorn (2003).

Substitution of (7) in the fraction of iron(II) dissolved (Fe^{2+}/Fe_T) equals:

$$\frac{[Fe^{2+}]}{[Fe]_T} = \frac{[Fe^{2+}]}{[Fe^{2+}] + [S{-}OFe(II)^+]} = \frac{1}{1+\frac{K_{Fe/H}[S{-}OH^0]}{[H^+]}} \approx \frac{1}{1+\frac{K_{Fe/H}[S^*]}{[(H^+]}} \tag{2.8}$$

Figure 2.4 shows the fraction of iron(II) dissolved as function of pH for two HFO concentrations. If Fe(II) is removed in sub soils or in a reactor, the concentration of HFO may be high. At a specific surface area of 600 m^2/g (Hiemstra and van Riemsdijk, 2009), 0.1 mol HFO/L represents a surface area of about 1500 m^2/L. For comparison, values of about 600–6000 m^2/L are typically for natural soils (Hiemstra et al., 2010). The large surface area is able to bind a very large part of the Fe(II) as illustrated in Figure 2.4, showing that for HFO concentrations >1 mol/L, less than 1% iron(II) is dissolved at pH values >6.5 to 7. This value corresponds well with experiences with subsoil iron removal.

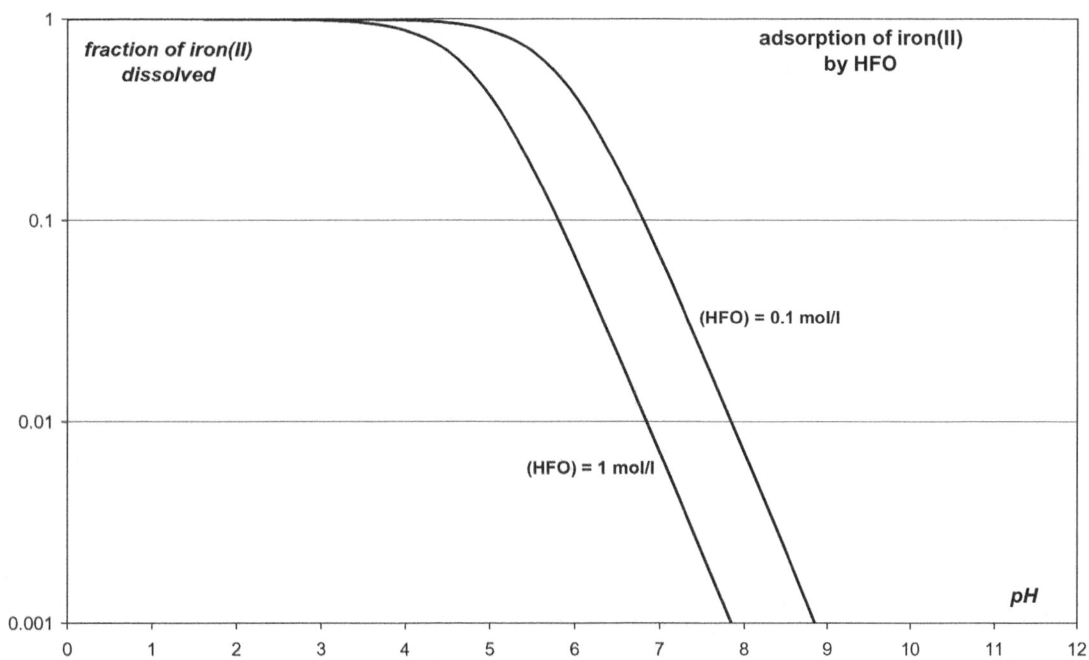

Figure 2.4 Fraction of iron(II) dissolved as function of pH for two representative HFO concentrations

Figure 2.3 also provides information about the adsorption rate: this rate must be faster than the rate corresponding with the minimum value of the half-life time at the same pH under conditions of simultaneous adsorption and oxidation.

2.3.3 Biological oxidation

The precipitation of iron(III) hydroxides by iron(II) and oxygen produces energy that may be used by micro-organisms. This energy production may be represented as:

Oxidation: $Fe^{2+} + \frac{1}{4}O_2 + H^+ \longrightarrow Fe^{3+} + \frac{1}{2}H_2O + x\ kJ/mol$

Hydrolysis: $Fe^{3+} + 3H_2O \longrightarrow Fe(OH)_3\downarrow + 3H^+ + y\ kJ/mol$

Sum: $Fe^{2+} + \frac{1}{4}O_2 + 2\frac{1}{2}H_2O \longrightarrow Fe(OH)_3\downarrow + 2H^+ + (x + y)\ kJ/mol$

The energy yield under actual conditions may be calculated as the Gibbs free energy. Figure 2.5 shows the energy yield of oxidation of iron(II) by oxygen as function of pH. This figure shows that the energy yield decreases with increasing pH, and that, under the represented conditions, at $pH > \sim 4$ energy should be supplied for this reaction. However, as the iron(II) concentration is too low (See Paragraph 2), this relation is not relevant at $pH > \sim 3$. Figure 2.5 also shows that at $pH < 2$ à 2.5, oxidation of iron(II) to iron(III) produces more energy than the combined reaction of oxidation and precipitation. Consequently, if no energy is added, at $pH < 2$ à 2.5, iron(II) is only oxidized to iron(III), and not precipitated as iron (III) hydroxide. Figure 2.5 also shows that higher pH values, higher iron(II) and higher oxygen concentrations favor the precipitation of iron(III) hydroxides, but that the energy yield is very dependent on pH and hardly on oxygen concentration.

Figure 2.5 shows that at $pH = 6$ the energy yield of oxidation and precipitation amounts to 50 à 60 kJ/mol. Others mention an energy yield of 40 à 50 kJ/mol (Chapelle, 1992). These values are for 25°C; at lower temperatures the energy yield will be somewhat less. For comparison: oxidation of organic matter by oxygen produces under representative groundwater conditions (0.1 μmol/L degradable organic carbon, 0.1 mmol/L oxygen and 1 mmol/L carbon dioxide) circa 450 kJ/mol. Consequently, oxidation of iron(II) yields per mol 8 to 10 times less energy than oxidation of organic matter.

Iron oxidizing bacteria (IOB), like *Gallionella ferruginea* and *Leptothrix ochracea*, are able to utilize the energy produced by oxidation and precipitation for their metabolism (Emerson, 2000; Ghiorse, 1984; Hallbeck and Pedersen, 1991; Hanert, 2002; Kappler and Straub, 2005; Konhauser, 1997). In order to obtain this energy, dissolved iron(II) should be captured first. IOB are able to adsorb iron(II) on their cell material. As an aid in the competition with heterogeneous oxidation, IOB excrete large amounts of EPS (Extracellular Polymeric Substance) (Ehrlich et al., 1991). These polymers possess a high affinity for iron(II) (Frankel and Bazylinski, 2003). Oxidation of iron(II) by enzymes on the cell wall (a.o. *Gallionella*) and extracellular by exopolymers (e.g. *Gallionella*, *Lepthotrix* and *Siderocapsa*) has been confirmed in several studies (e.g. Chan et al., 2009; Bourgine et al., 1994; Czekalla et al., 1985; Cullimore and McCann, 1978).

Because the precipitation of iron(III) hydroxides from iron(II) and oxygen produces only little energy, IOB need to convert large quantities of iron(II) (Chapelle, 1992; Viswanathan and Boettcher, 1991; Ghiorse, 1984). This is only possible where both oxygen and iron(II) are present. For that reason, IOB are typical representatives for oxygen–iron(II) interfaces (Emerson and Weiss, 2004), characterized by variable iron(II) but low oxygen concentrations and slightly acid to neutral ($5.5 < pH < 7.5$) conditions.

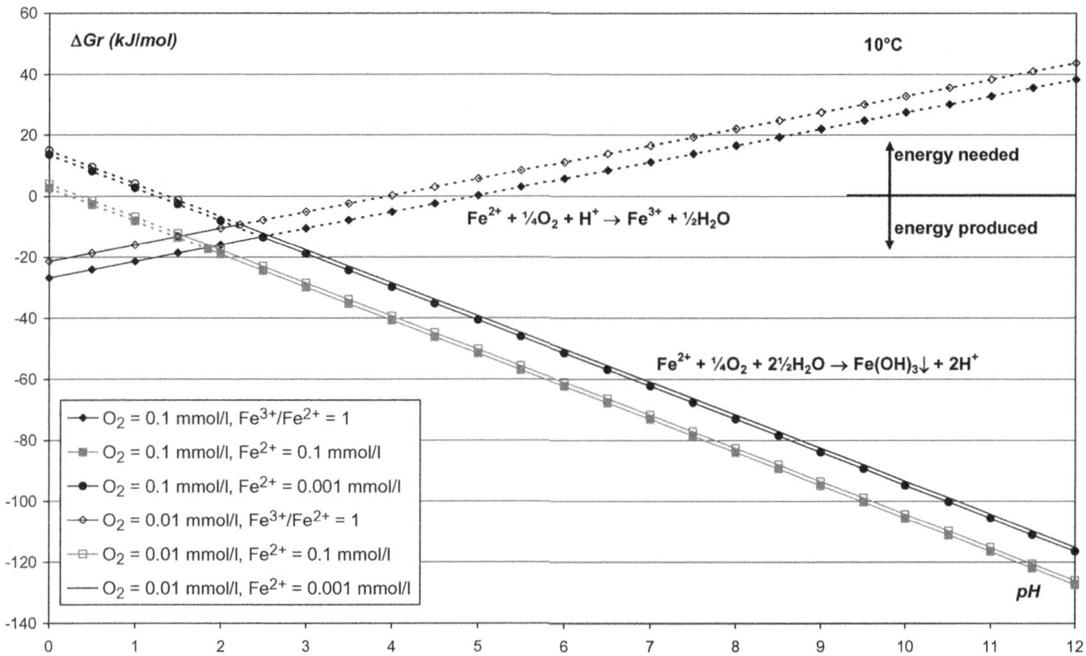

Figure 2.5 Energy yield for oxidation of iron(II) by oxygen to iron(III) and of precipitation of HFO as function of pH for various concentrations of oxygen and iron(II) and for some iron(III)/iron(II) ratios. The relevant energy yield is represented by the full drawn lines for $O_2 = 0.1$ mmol/l (closed symbols), $O_2 = 0.01$ mmol/l (open symbols) and iron(II) = 0.001 mmol/l (upper line). The energy yield for iron(II) = 0.1 mmol/l (lower line) is in analogy

2.3.4 Synthesis

Figure 2.6 presents a summary of the conditions where the various oxidation processes are dominant. In this figure, data were used as presented in literature, which may refer to different kinds of iron(III) hydroxides, and no interferences were considered, such as complexation of iron(II) with for instance organic matter. This figure shows that homogeneous iron(II) oxidation is only relevant in the absence of HFO. As soon as HFO are present, heterogeneous oxidation is the dominant process. Homogeneous and heterogeneous iron(II) oxidation are equal to each other if both rates are equal:

$$k_2 \frac{[S^*][Fe^{2+}][O_2]}{[H^+]} = k_1 \frac{[Fe^{2+}][O_2]}{[H^+]} \Rightarrow [S^*] = \frac{k_1}{k_2}\frac{1}{[H^+]} \tag{2.9}$$

As soon as IOB are present, competition starts between biological and heterogeneous iron(II) oxidation. This competition is not so much a competition for oxygen but much more for iron(II). And not the oxidation rate of iron(II) is important, but much more the iron(II) adsorption rate: adsorption by IOB versus adsorption by HFO (Böhler, 1999; Ehrlich et al., 1991; Cullimore and McCann, 1978). Because, if IOB are not successful in capturing iron(II), the energy produced by precipitation can not be to their advantage.

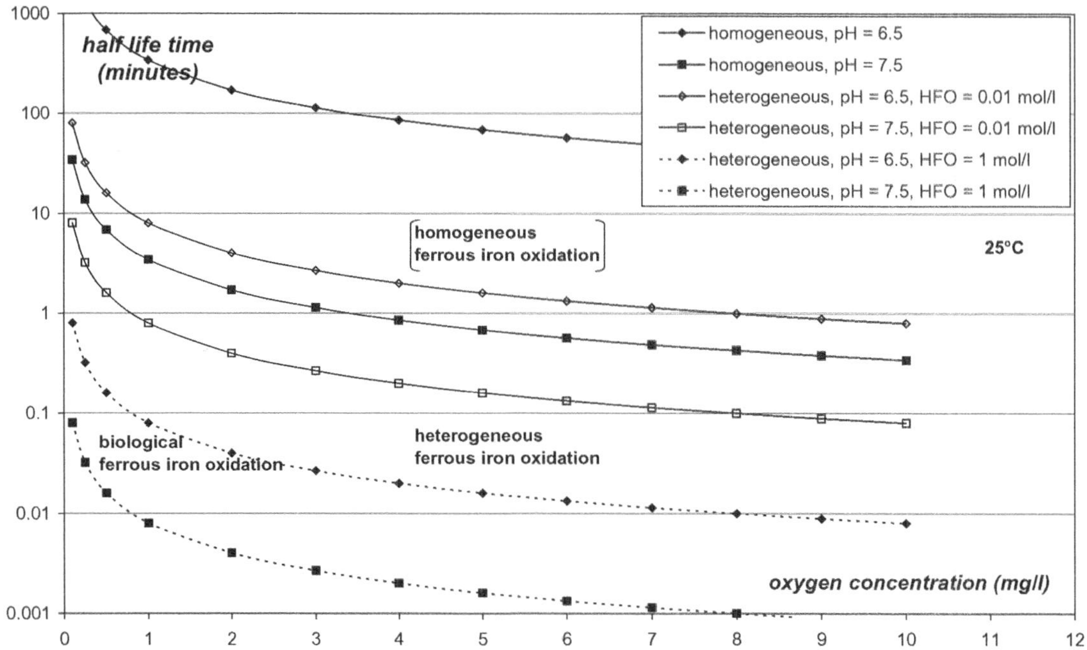

Figure 2.6 Conditions where homogeneous, heterogeneous or biological oxidation are relevant. Homogeneous oxidation is only relevant in the absence of heterogeneous oxidation. The "biological" field is only approximate, relevant at low pH and low oxygen concentration

2.4 BIOLOGICAL IRON REMOVAL OR HETEROGENEOUS AND BIOLOGICAL IRON OXIDATION?

Biological iron removal is considered by many scientists as the main process of Fe(II) removal, in particular referring to Mouchet (1992) and Czekalla (1988). Mouchet (1992) remarked that various conventional iron removal plants operated satisfactorily even though this was not expected from the chemical composition of the abstracted groundwater. Because microscopic examination of sludge samples revealed massive growth of IOB, he ascribed this favorable removal to biological activity. Moreover, there were no problems with silica interference and filter run times were longer than in conventional plants. Mouchet (1992) mentioned still other advantages of biological oxidation: formation of a more compact iron sludge, high retention capacity, high filtration rates and insensitivity to discontinuous operation. Moreover, prolonged shutdowns do not negatively affect treatment results (Czekalla and Kotulla, 1990), nor do variations in water composition (Groth and Czekalla, 2004). In applying this method, the oxygen concentration should not exceed the concentration necessary for oxidation of iron(II). Moreover toxic constituents like H_2S, heavy metals and hydrocarbons should be absent.

Hässelbarth and Lüdemann (1971) were (one of) the first mentioning biological iron removal. In their research on chemical well clogging (clogging of the screen slots by iron(III) hydroxide precipitates and biomass), they mentioned that great amounts of these precipitates and IOB were accompanied by low oxygen concentrations. Apparently, they considered this a causal relationship, and reversed the reasoning: by reducing the oxygen concentration, to 0.02 mg/L, IOB would develop profusely and remove dissolved iron(II) from groundwater.

Czekalla et al. (1985) and Czekalla (1988, 2008) listed the IOB present in groundwater treatment plants, and noted that iron was only oxidized in direct contact with IOB. They distinguished two mechanisms in iron (II) oxidation: intracellular oxidation and extracellular oxidation with the help of EPS. Later, Czekalla (1997) successfully applied biological iron removal with higher oxygen concentrations (4 to 11 mg/L), which was only possible as long as the residence time between aeration and filtration was short in order to prevent premature (homogeneous) iron(II) oxidation.

However the above phenomena, attributed to biological iron(II) removal, are also characteristic for heterogeneous oxidation.

Oxidation processes proceed along a standard sequence (Stumm and Morgan, 1996; Appelo and Postma, 2006), i.e. $CH_4 > H_2S > Fe^{2+} > NH^{4+} > Mn^{2+}$. Consequently, one would expect that CH_4 and H_2S are oxidized before iron(II). However, Grohmann et al. (1989) and Böhler (1999) mention that methane does not disturb biological iron removal. They attribute this phenomenon to the quantity of oxygen which is less than equivalent for complete iron(II) oxidation: $(O_2) < 0.14(Fe^{2+})$, in mg/L. This observation may be more conveniently explained by a difference in processes: iron(II) being removed by adsorption and CH_4 by oxidation. De Vet et al. (2002) confirmed the absence of interference of methane with iron in iron removal filters.

Based upon the density of the iron sludge and the characteristics of the abstracted groundwater, such as redoxpotential, pH and oxygen concentration, Søgaard et al. (2001, 2000) classify their groundwater treatment plants as either biological or chemical. They classify water treatment plant Forum (pH = 5.5) as chemical, and the plants Astrup and Grindsted (pH = 7.2 to 7.3) as biological. Considering pH one would expect just the opposite classification.

Considering the results of the discussion concerning iron(III) hydroxide precipitation, biological iron removal is evidently not identical to biological iron oxidation. This conclusion is confirmed by results of Tremblay et al. (1998). After backwashing a biological iron removal filter, they could not detect any IOB in the sludge. Nevertheless, despite the absence of IOB, they still called the process biological iron removal.

Evidently, what in literature often is called biological oxidation, actually refers to a process technological method, characterized by high filter velocities and limited oxygen supply. In order to limit confusion as much as possible, this method will be called here biological iron(II) removal. The expression biological iron(II) oxidation is reserved here for the process of oxidation of iron(II) by IOB. Consequently the method of biological iron removal actually refers to the process of heterogeneous iron(II) oxidation, whether or not together with biological iron(II) oxidation.

2.5 DISCUSSION AND RECOMMENDATIONS

The distinction between the method of biological iron(II) removal and the process of biological iron(II) oxidation leads for some observations to alternative views. According to Czekalla (1988) and Mouchet (1992), chemical iron(II) removal changes into biological removal by restricted aeration, because lower oxygen concentrations result in more favorable conditions for *Gallionella*. However, the contribution of heterogeneous and of biological iron(II) oxidation to iron(II) removal is governed by the adsorption rate, which is independent of the oxygen concentration. The favorable effect of lowering the oxygen concentration (longer runtimes) consists of minimizing homogeneous oxidation (and in this way indirectly favoring heterogeneous and biological oxidation).

Regular backwashing the filters with aerated water has a favorable effect on biological iron(II) removal (Bourgine et al., 1994). In biological iron removal, the oxygen supply is critically adjusted to the iron (II) concentration, which may lead to incomplete oxidation of adsorbed iron(II) in (a part of) the filter. Backwashing with aerated water then results in a thorough oxidation of the adsorbed iron(II). Moreover, backwashing is also necessary in order to separate the growing iron(III) hydroxide pellets from each other.

By backwashing, more or less IOB biomass will be removed and, by consequence, thereafter the contribution of biological oxidation to iron removal will be less. Over time the IOB biomass will increase, and concurrently the contribution of biological oxidation until the next back washing, after which the cycle starts over again. Thus, back washing results in a cyclic variable contribution of biological oxidation to iron(II) removal.

Due to continuous deposition of iron(III) hydroxide by heterogeneous oxidation, the size of the iron(III) hydroxide pellets will increase. This increase in size results in an increase of the height of the filter bed (Frischherz et al., 1985) and in a decrease in specific surface area, necessitating, in regular intervals, the replacement of large pellets by seeds. This increase in bed height over time may be used to estimate the contribution of heterogeneous iron(II) oxidation to total oxidation.

In filters fed by oxygen free groundwater, consequently containing no IOB, no IOB will be present (Bruins et al., 1998), unless they have been inoculated. In The Netherlands the occurrence of oxygen free groundwater is limited to (semi-)confined aquifers. The pH of these abstracted groundwaters varies roughly between 6.8 and 7.6, making them excellently suited for iron(II) removal by heterogeneous oxidation (see also Sharma et al., 2001).

In filters fed by groundwater containing oxygen and iron, IOB will always be present, as continuously delivered by the abstracted groundwater. In The Netherlands the occurrence of oxygen and iron containing groundwater is limited to phreatic aquifers. The pH of these abstracted groundwaters is somewhat lower. Well fields abstracting iron and oxygen containing water are characterized by the occurrence of screen slot (chemical) well clogging. The premature iron(III) hydroxide precipitation will also clog field (raw water) lines, and interfere with the treatment to drinking water. In these filters iron (II) will be removed by heterogeneous and by biological oxidation.

After aeration of abstracted groundwater, homogeneous iron(II) oxidation will occur in the supernatant water of the filters. In order to prevent clogging of the filter surface as much as possible, this volume of supernatant water should be minimal.

The results of this review demonstrate that the dominant process in iron(II) removal consists of heterogeneous oxidation (95 to 98% of the total iron(II) oxidation?), with a negligible contribution of homogeneous oxidation (2 to 5%?), see also Teunissen et al. (2007). If biological oxidation occurs, this process will add a variable, and pH-dependent, contribution (up to 25%?) at the expense of heterogeneous oxidation.

Efficient iron removal asks for creating optimal conditions for iron(II) adsorption. Homogeneous and biological iron(II) oxidation must be restrained as much as possible, as they increase filter resistance, making back washing necessary. In heterogeneous iron(II) oxidation, the iron(III) hydroxide grains will grow continuously by precipitation of fresh material. In order to prevent cementing of these grains, and in order to achieve complete oxidation of adsorbed iron(II), periodic upflow flushing of the filter is necessary. This periodic flushing may be omitted by passing the abstracted groundwater through an upflow pellet reactor filled with iron(III) hydroxide pellets, whether or not with continuous or discontinuous oxidation. Advantage of an upflow reactor is a more intimate contact between liquid and surface, a disadvantage might be rubbing of the pellets against each other by which the fresh, most reactive, surfaces are removed continuously.

Optimalization of iron(II) removal by heterogeneous oxidation is quite possible on plants equipped with double filtration (de Ridder et al., 2008), which abstract oxygen free groundwater with a high pH. Actually subsoil iron removal is performed above ground (de Vet et al., 2009; Sharma, 2001; van Beek, 1983).

Koreman and Heuvingh (1997, 1998) did already semi-field experiments with an "upflow pellet reactor". Initially, these experiments were successful, but after some time the removal efficiency decreased by unknown reason. Dietz and Dempsey (2002) and Diz and Novak (1998) applied similar reactors for removal of iron(II) in Acid Mine Drainage (AMD).

2.6 CONCLUSIONS

Three processes compete for the oxidation and precipitation of iron(II): homogeneous, heterogeneous and biological oxidation. The distribution between these processes is governed amongst others by pH, oxygen concentration, iron(III) hydroxide and IOB, including EPS, concentration and adsorption rates.

Biological iron removal, characterized by limited oxygen supply and high filtration rate, refers to a process technological (operational) method, and must not be confused with the process of biological iron oxidation. Biological iron removal consists of heterogeneous iron(II) oxidation, whether or not together with biological iron(II) oxidation.

In order to present an indication on the distribution of the various processes for iron removal from groundwater in drinking water treatment, it is estimated that iron(II) removal will be up to 5% by homogeneous, for 75 to 100% by heterogeneous, and for 0 to 25% by biological oxidation. If present, the contribution of biological oxidation is not constant over time: after backwashing the contribution by IOB will be lowered, and will increase over time at the expense of heterogeneous iron oxidation untill the next backwashing.

For drinking water treatment it would be advantageous to design an installation directed towards optimal removal of iron(II) by heterogeneous oxidation.

Chapter 3

Concentration and size distribution of particles in abstracted groundwater*

C.G.E.M. van Beek[1], A.H. de Zwart[2], M. Balemans[1], J.W. Kooiman[1], C. van Rosmalen[3], H. Timmer[4], J. Vandersluys[5,†] and P.J. Stuyfzand[1,6]

[1]*KWR Watercycle Research Institute (formerly Kiwa Water Research), P.O. Box 1072, 3430 BB Nieuwegein, the Netherlands; +31 30 60 69 511; fax +31 30 60 61 165; Kees.van.Beek@kwrwater.nl.*

[2]*Technical University Delft, now Shell Research and Exploration, Rijswijk, the Netherlands*

[3]*Water utility Brabant Water*

[4]*Water utility Oasen*

[5]*Water utility Vlaamse Maatschappij voor Watervoorziening, VMW*

[6]*VU University, Amsterdam*

ABSTRACT

Particle number concentrations have been counted and particle size distributions calculated in groundwater derived by abstraction wells. Both concentration and size distribution are governed by the discharge rate: the higher this rate the higher the concentration and the higher the proportion of larger particles. However, the particle concentration in groundwater derived from abstraction wells, with high groundwater flow velocities, is much lower than in groundwater from monitor wells, with minimal flow velocities. This inconsistency points to exhaustion of the particle supply in the aquifer around wells due to groundwater abstraction for many years. The particle size distribution can be described with the help of a power law or Pareto distribution. Comparing the measured particle size distribution with the Pareto distribution shows that particles with a diameter >7 μm are under-represented. As the particle size distribution is dependent on the flow velocity, so is the value of the "Pareto" slope β.

*Published slightly modified as: C.G.E.M. van Beek, A.H. de Zwart, M. Balemans, J.W. Kooiman, C. van Rosmalen, H. Timmer, J. Vandersluys and P.J. Stuyfzand (2010): Concentration and size distribution of particles in abstracted groundwater. Water Research 44 (3) 868-878, doi: 10.1016/j.watres.2009.09.045.

3.1 INTRODUCTION

Groundwater contains solids and biomass (McDowell-Boyer et al. 1986), which has been linked with transport of pollutants (McCarty and Zachara 1989) like pesticides and radio active elements, and of pathogenic bacteria and viruses (Schijven 2001; Foppen 2007).

The concentration of solids in groundwater has most frequently been measured by filtering groundwater over membrane filters with successively smaller diameters (McCarthy and Degueldre 1993), yielding particle mass concentrations (mg/l), and much less frequently with particle counters (Hofmann 1998; Dehnert et al. 2003; Marre 2004), yielding particle number concentrations (n/ml). These measurements were executed with the least possible disturbance during sampling.

In this contribution we present information on the presence and behavior of particles in groundwater abstracted by wells during normal operation, i.e. frequent switching on and off. Reason for our particle count study was the suspected role of particles in well bore clogging (van Beek 2002), which has been ascertained by Timmer et al. (2003) and de Zwart (2007). The cause and prevention of well bore clogging by particles has been described elsewhere (Chapter 5).

The results of our research are not only important for understanding the transport of particles as carriers of pollutants and pathogenic micro organisms, but also for related fields, such as artificial recharge (Huisman and Olsthoorn 1983), artificial storage and recovery (ASR), aquifer thermal energy storage (ATES) and water flooding in oil production (Vaidya and Fogler 1990), and for construction of wells (Saucier 1974; Gruesbeck and Collins 1982), not only for drinking water production, but also for irrigation, dewatering of building excavations (Reddi 1997; Powers 1992) and pressure relief of dams (ACE 1992).

3.2 MATERIALS AND METHODS

3.2.1 Particle counter

In our research particles were counted using a Met One online particle counter, equipped with a PCX sensor. This sensor is able to count particles with a diameter between 2 and 750 μm. This upper diameter is not realistic, since in the field the upper diameter is limited by the diameter of the pore throats between the grains of the aquifer matrix. At the sites studied, the upper particle diameter varied between 25 to 50 μm. Consequently, in this study particles are defined as inorganic and organic solids with an equivalent diameter >2 μm, where the upper diameter limit has not been specified.

The operation of the PCX sensor is based on the light-blocking method (Hargesheimer and Lewis 1995): a laser beam is directed through a flow-through cell onto a light-sensitive electrical sensor at the opposite side. Particles crossing this laser beam project a shadow onto this electrical cell. The number of shadows equals the number of particles, and from the magnitude of the shadow, an equivalent diameter of the passing particle is calculated by an internal code. The results are presented in six user defined diameter classes. These classes are defined in our experiments as 2–3, 3–5, 5–7, 7–10, 10–15 and >15 μm.

The particle concentration can be expressed as number, volume or mass of particles per unit volume (Lerman 1979). The particle counter only counts the numbers of particles, so we have presented the particle concentration as the particle number concentration (number of particles present per unit volume), and not otherwise, as these latter representations require assumptions about shape and density.

The Met One particle counter operates online. The concentration is calculated by dividing the number of particles counted during a specified time interval by the volume flow rate. The flow rate used was

100 ml/minute, as advised by the manufacturer. This flow was checked regularly, as a deviation in this flow rate results in a corresponding deviation in particle concentration. The counting time interval was set at 1 min. This time interval resulted in smooth concentration curves over time, but can still indicate short-term variations.

To assure the quality of the measurements, the particle counter was regularly calibrated by the supplier with the help of latex spheres. These spheres are different from particles present in groundwater. Particles in groundwater may have all kind of shapes, such as spheres, platelets and needles, and a range of densities, such as solids, flocs and cell biomass. These different shapes imply that the position of the particle in the laser beam is also important: a platelet may project itself full face as a rounded particle and on its side as a needle. Moreover, the transparency of particles for light may vary from transparent to opaque, and may also vary over the particle. Due to this variation in position and transparency, the results of particle counters do not represent absolute values, but since all measurements were conducted identically, the results are mutually comparable.

For our research, two identical counters were available. Parallel counting by splitting the sample flow showed that both counters did not yield identical results, despite the periodic calibration. This difference specifically concerned the particle concentration and less the particle size distribution, again providing relative instead of absolute results. In order to minimize this deficiency, most counting was done with a single counter.

Since August 2005, the measurements were conducted with the help of a PAMAS Waterviewer, which works on the same principle and has the same deficiency.

3.2.2 Geology, wells and groundwater quality

The geology of the Netherlands mainly consists of unconsolidated sedimentary deposits, varying in thickness from zero in the SE to more than 500 m in the NW. Groundwater abstraction occurs between 20 to 250 meter below surface (mbs). Because its inertness, nearly all wells involved were made from PVC. All wells are equipped with submersible pumps. Depending on the size of the well, the capacity of these pumps ranged between 40 and 100 m^3/h. These pumps are kept in operation until malfunction, after which they are replaced.

The particle counter was connected to the sample valve on top of the well with a plastic (PE) tube. However PE is permeable to oxygen, and iron-hydroxide flocs may be generated, the use of this tube was preferred because it is very flexible. In order to prevent possible interference, the tube was replaced each 1 to 3 days.

Before counting, the well to be measured was transferred from automatic mode to manual operation. Manual operation ensures that the well was running continuously several hours before and during particle counting and that the results are representative for equilibrium abstraction conditions. During this same time period, the sample valve, which is mounted on the well head, was turned wide open. In this way the flowing water would scour off any particles present in its body.

Around non clogging and clogging wells particles accumulate on the gravel pack/aquifer interface during abstraction; in non clogging wells all these accumulated particles are removed by the next switching on, in clogging wells not all particles are removed (Chapter 5). As particles in abstracted groundwater are subject to similar processes, the results of all counts have been put together.

All sampled wells abstract fresh groundwater for the public drinking water supply. In order to ensure that all counted particles indeed originated from the aquifer, our research was limited to wells abstracting anaerobic (sulfate reducing) groundwater. Because the abstracted groundwater contains no oxygen, well screen clogging by accumulation of iron hydroxides, manganese oxides, and/or accompanying biomass is impossible (van Beek 1984). Consequently, the results of our particle counts cannot be biased by these precipitates.

However, abstracted groundwater from all wells contains methane, ranging from <0.01 to several mg/l. Dissolved methane may develop into gas bubbles, causing complications because the refraction of light at the liquid-gas interface may present itself as a shadow. Fortunately, the presence of gas bubbles is shown as an overrepresentation of large particles.

Oxygen entry into stagnant water in the well is negligible, since diffusion of gasses across the water-atmosphere boundary is very slow. After passing the submersible pump, the abstracted groundwater remains under pressure, which excludes entry of oxygen as long as the joints are airtight. The absence of iron hydroxides is confirmed by the lack of fouling of the flow-through cell of the particle counter.

3.3 RESULTS AND DISCUSSION

3.3.1 Particle concentration

Figure 3.1 shows the results of particle counts in groundwater abstracted from well 26 on well field Tull en 't Waal (water utility Vitens Midden Nederland) during one week. The concentration of particles with a diameter >2 μm ranges from about 2 to almost 2000/ml during peaks. Groundwater abstracted from well 26 is, together with other wells, transported by a collector waterline to the treatment plant. In this plant the abstracted groundwater is processed to drinking water by removing iron, manganese and ammonium by aeration and filtration. In order to remove the accumulated iron, manganese and biomass, the rapid sand filters are backwashed between 3 and 4 AM. For this backwashing, it is necessary to switch the submersible pump on and off several times. Apparently, switching on the pump results in very high peak concentrations. During the day (3 PM), when other wells are also abstracting and delivering to the same collector water line, the discharge rate of well 26 is somewhat lower than during the night (12 PM). During the night, well 26 is the only well abstracting, resulting in a lower resistance of the collector water line, and consequently in an increased discharge rate. This fluctuating discharge rate results in fluctuating particle concentrations. This is especially apparent during the weekend of 11 and 12 August 2002, when there was no backwashing.

The dependence of the particle concentration on the discharge rate is shown in greater detail in Figure 3.2. Starting the submersible pump results in a high initial concentration, which gradually levels off to a nearly constant value. The same leveling off happens after each change in the discharge rate of the pump. Note the difference in behavior of the particle concentration after decreasing and after increasing the discharge rate. Apparently, the acceleration introduced by increasing the discharge rate results into a greater disturbance around the well than the deceleration by decreasing the discharge rate. Although the concentration has not reached an equilibrium value, it is clear that this equilibrium value depends on the magnitude of the discharge rate, i.e., the higher the discharge rate, the higher the concentration.

Figure 3.3 shows the cumulative distribution of the particle concentration, measured in 62 wells. All measurements were conducted under normal operating conditions of the well fields. The results are presented as measured, despite variations in aquifer properties and in well construction.

The data points in Figure 3.3 do not represent calculated averages including peak concentrations, but correspond with nearly constant concentrations during lengthy abstraction. For example, as the normal discharge rate of well 50 in Figure 3.2 equals 65 m^3/h, the concentration of particles >2 μm is estimated at about 185/ml. The cumulative distribution diagram shows that the median concentration of particles with a diameter >2 μm in the abstracted groundwater equals 9/ml and that 90% of all values range between 2 and 120/ml, with extremes from 0.6 to 340/ml.

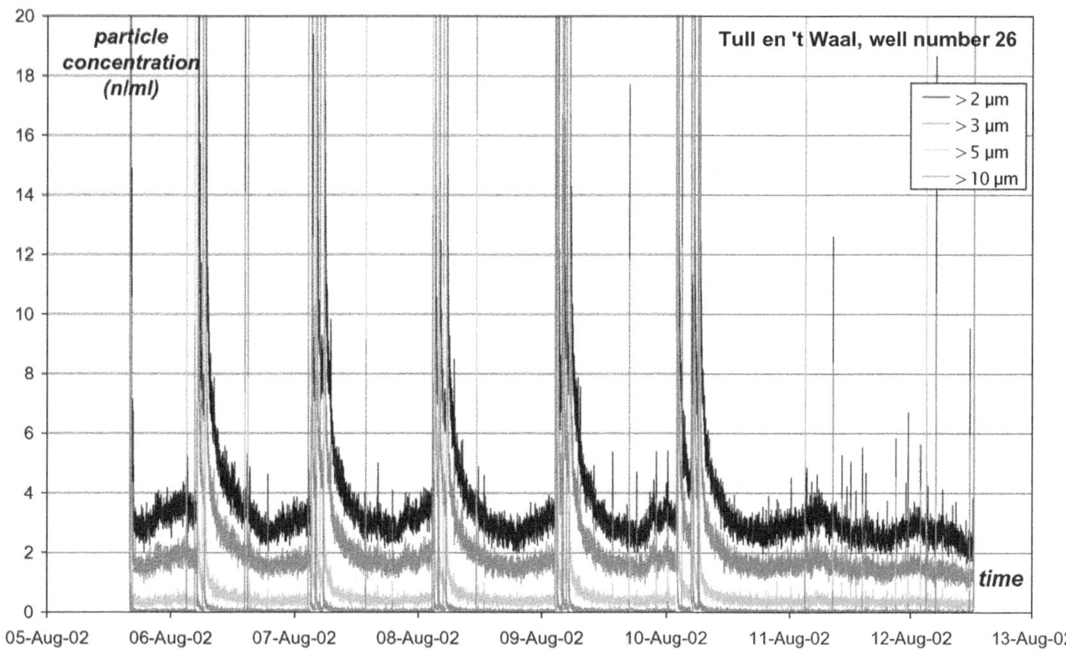

Figure 3.1 Concentration of particles in abstracted groundwater from well 26 of well field Tull en't Waal (water utility Vitens Midden Nederland). The concentration has been averaged over each minute; this series consists of $7 * 24 * 60 = 10{,}080$ measurements

3.3.2 Particle size distribution

Figures 3.1 and 3.2 showed the cumulative particle concentration in abstracted groundwater. These figures also contain information about the particle size distribution. However, it is more convenient to present this information in a separate graph (Figure 3.4). This graph shows the cumulative particle size distribution over time, where the fraction of particles >2 µm equals 1 (all particles measured are >2 µm).

Figure 3.4 shows that also the particle size distribution is influenced by the abstraction rate: the higher the discharge rate, the greater the relative and absolute contribution of larger particles. Depending on the abstraction rate, the fraction of particles with a diameter >3 µm varies between 0.46 and 0.57, with a diameter >5 µm between 0.09 and 0.13, and with a diameter >10 µm between 0.04 and 0.07.

Figure 3.5 shows the cumulative distribution of the particle size fractions with a diameter >3, >5 and >10 µm, respectively. In this figure the fractions of particles >3 µm are presented in ascending order and vary between 0.45 and 0.67, with a median value of 0.57. In the abstracted water from the outlier value, gas bubbles were observed during particle counting.

In Figure 3.5 the particle size fractions >5 µm and >10 µm are presented in corresponding order with the fraction >3 µm. These particle size fractions >5 and >10 µm run more or less parallel to the fraction >3 µm: smaller fractions >3 µm correspond with smaller fractions >5 and >10 µm, and vice versa. Inspection showed that there is no general relation between concentration of particles >2 µm and particle size fractions.

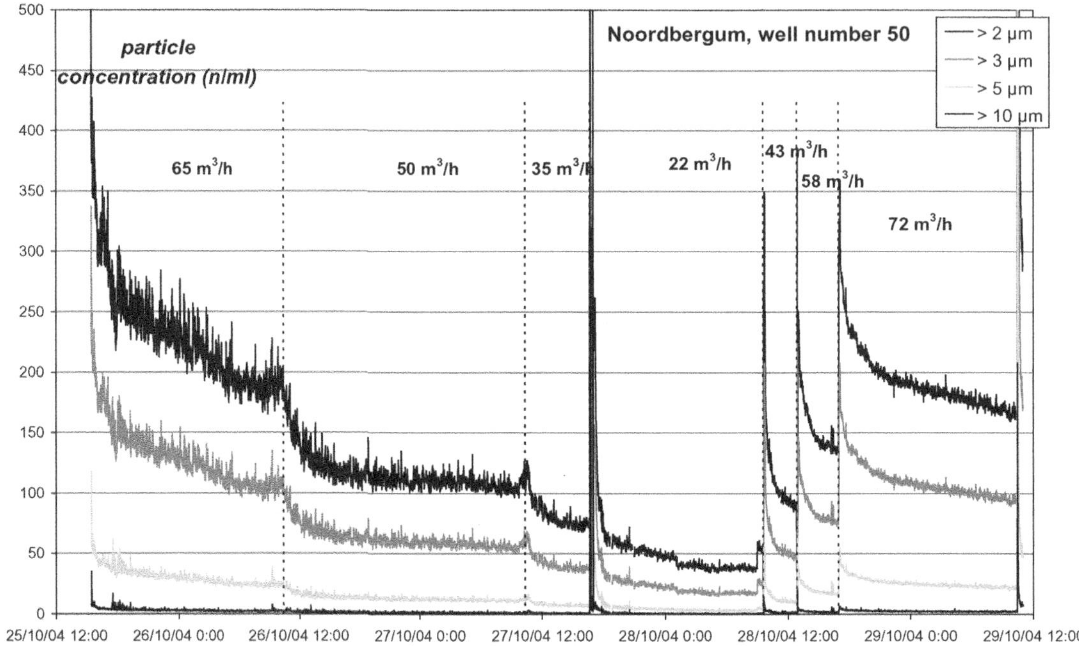

Figure 3.2 Particle concentration during continuous abstraction with variable discharge rate from well 50 of well field Noordbergum (water utility Vitens Friesland). The anomalous peak in particle concentration during the change in flow rate from 35 to 22 m^3/h was caused by temporarily insufficient water supply to the particle counter

3.3.3 Power law or Pareto distribution

Many particle size distributions (sediments, particles in air and water) may be described by an empirical relationship, i.e. the power law or Pareto distribution (Lerman 1979). This distribution is expressed as:

$$\frac{dN}{d\varphi_p} = A{\varphi_p}^{-\beta} \tag{3.1}$$

with:

N : particle concentration (n/ml)
φ_p : particle diameter (µm)
A, β : constants

Introducing differentials and changing to logarithms yields:

$$\log \frac{\Delta N}{\Delta \varphi_p} = \log A - \beta \log \varphi_p \tag{3.2}$$

This relationship is often presented graphically, where the value for φ_p is obtained as $\sqrt{(\varphi_1 \varphi_2)}$, where φ_1 and φ_2 are the lower and upper boundaries of the particle diameter interval considered.

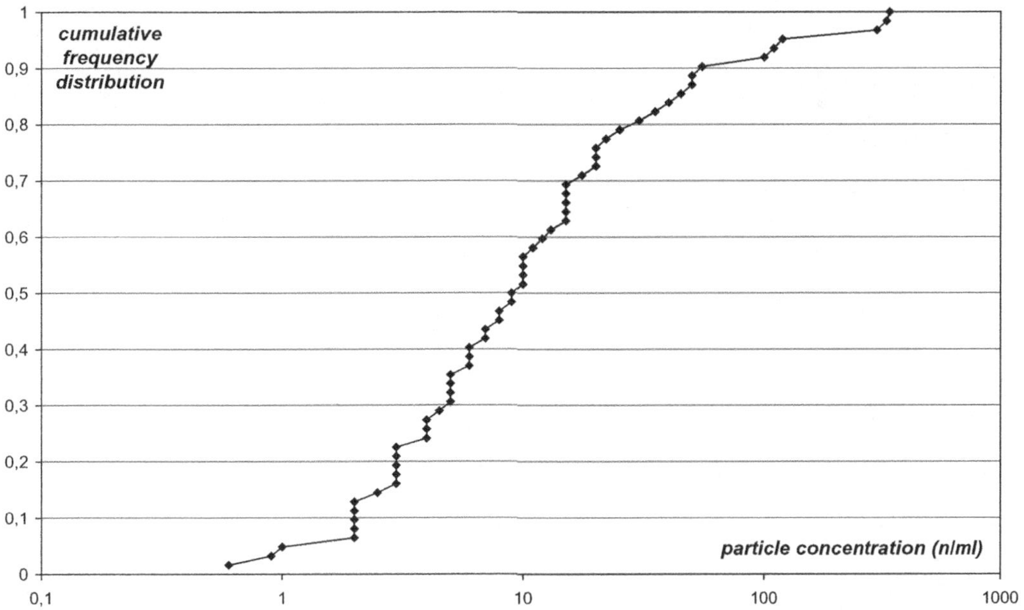

Figure 3.3 Cumulative frequency distribution of the concentration of particles >2 µm in abstracted groundwater under normal operating conditions. Each point represents a well

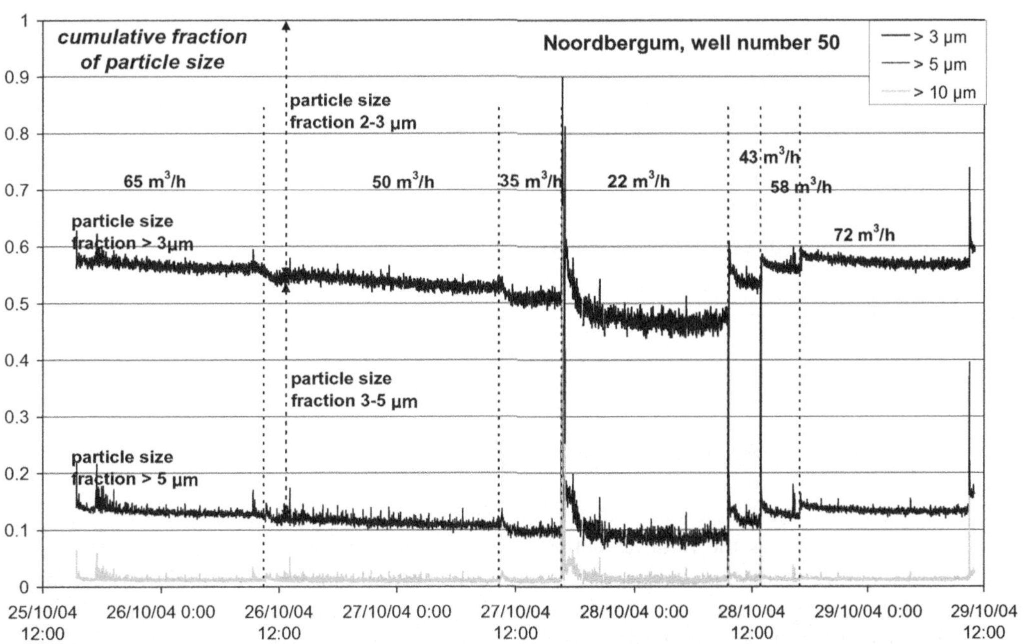

Figure 3.4 Cumulative particle size distribution during continuous abstraction with variable discharge rates for well 50 of well field Noordbergum (water utility Vitens Friesland). Compare with Figure 3.2

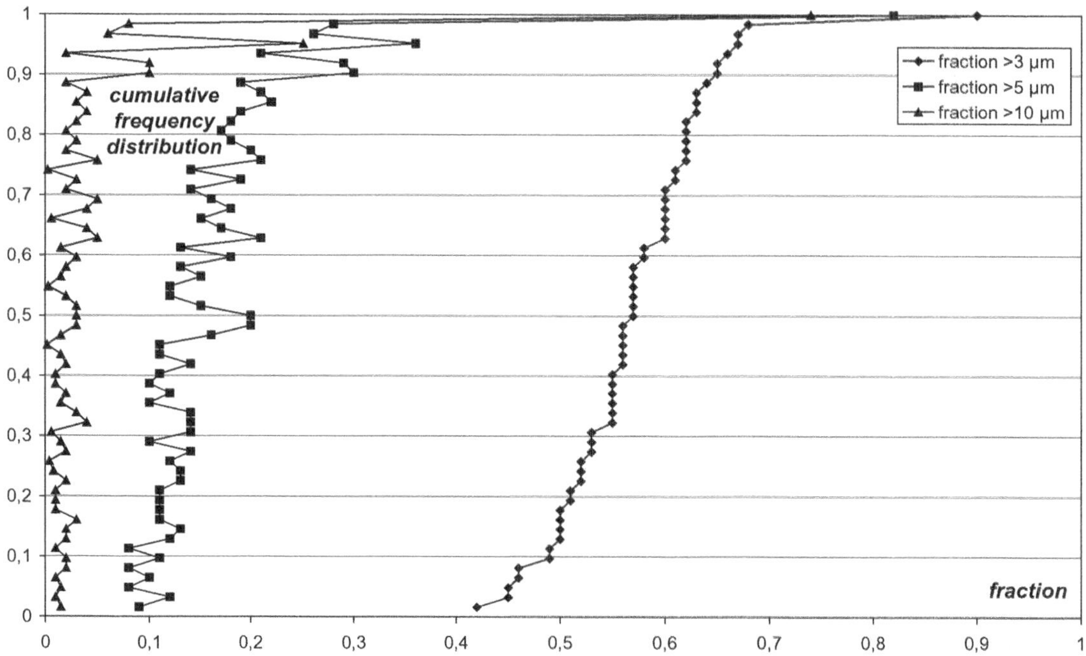

Figure 3.5 Cumulative frequency distribution of the fractions of particle concentrations with diameters >3, >5 and >10 μm respectively

Integrating Eq. (3.1) yields for $\beta > 1$:

$$N_{\varphi>\varphi_p} = \int_{\varphi=\varphi_p}^{\varphi=\infty} A{\varphi_p}^{-\beta} d\varphi_p = \frac{A}{\beta - 1}\varphi_p^{-(\beta-1)}$$

Changing to logarithms, this distribution reads:

$$\log N_{\varphi>\varphi_p} = \log\frac{A}{\beta - 1} - (\beta - 1)\log\varphi_p \tag{3.3}$$

with:

$N_{\varphi>\varphi p}$: concentration of particles with a diameter $\varphi > \varphi_p$

Equation (3.3) shows that the Pareto distribution may be characterized by one value for the concentration and slope β. Figure 3.6 shows the relation between particle concentration, calculated as the average of the last 20 counts before changing the discharge rate, and particle diameter on double logarithmic scale for well 50 of well field Noordbergum (Figures 3.2 and 3.4). According to Eq. (3.3), this relation should be linear with slope -(β – 1). The relation appears indeed linear between 2 and 7 μm, but the values for 3 μm deviate systematically (due to improper calibration of this diameter). Apparently, the value should read about 2.8 μm.

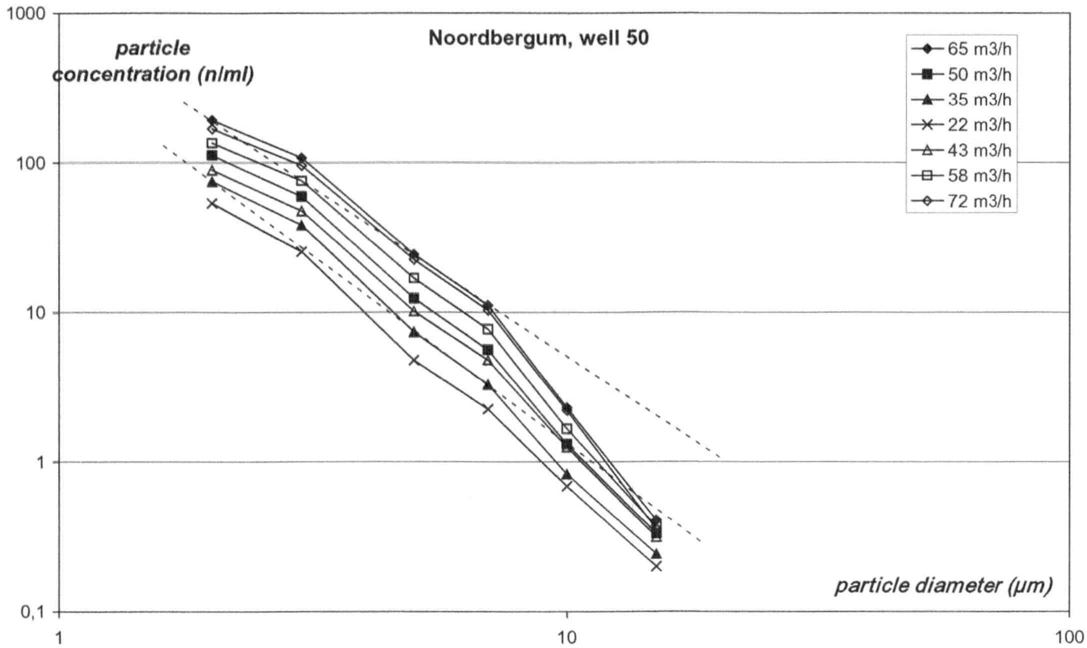

Figure 3.6 Double logarithmic plot of the particle concentration against the particle diameter for various discharge rates for well 50 of well field Noordbergum (water utility Vitens Friesland). The dotted lines represent the slopes for 65 and 35 m^3/h. Note that the order in the legend reflects the sequence of consecutive pumping rates

Figure 3.6 shows another interesting phenomenon: the concentration of particles with a diameter >7 µm is underrepresented. This indicates that either the supply of these particles falls short or the diameter of the pore throats at the well bore is so small that these particles cannot pass.

The grain size of the aquifer imposes an upper limit on the diameter of particles able to pass the aquifer matrix. In the tightest structure, the maximum diameter of a particle (d_p), which can pass a pore throat equals about 1/6 of the diameter of the grains (D_g): $d_p = D_g(2 - \sqrt{3})/\sqrt{3} = 0.155D_g$. Alluvial sediments show a stratification of alternating finer and coarser layers. Larger particles may pass coarser layers, but not finer ones. A well screen crosses these alternating finer and coarser layers in vertical direction. Thus the particle distribution measured at the top of the well actually consists of a sum of distributions, which are cut off at variable diameters on the coarse side. This results in an under-representation in the mixture of particles with a diameter larger than this cut-off value, and consequently in a deviating steeper slope in the Pareto diagram (Figure 3.6).

Moreover, the diameter of the pore throats shows a minimum at the aquifer/gravel pack (well bore) interface (van Beek 2002; Chapter 5). This minimum is caused by fine sand, carried by the drilling mud, trapped between gravel pack and aquifer, and/or remnants of drilling mud not removed during development. Bradford et al. (2005) showed experimentally that straining at textural interfaces from fine (aquifer) to coarse (gravel pack) is negligible, but relevant from coarse to fine.

Finally, the grain size of local unconsolidated sedimentary aquifers operated for groundwater abstraction ranges between around 300 and 500 µm. Using the value of 0.155 to calculate the maximum particle diameter able to pass the corresponding pore throats results in maximum particle diameters of 50 to 80 µm. This diameter is larger than the maximum diameters of 20 to 50 µm found in particle counts and the diameter corresponding with the break in slope in the Pareto diagram, i.e. 7 μm. This inconsistency may be explained by the formation of particle bridges, where pore throats are blocked by a number of particles with a diameter smaller than the corresponding throat diameter. According to Sen and Khilar (2006) bridging occurs at d_p/pore throat diameter ratios between 0.1 and 0.6. Converting pore throat diameter into soil matrix grain diameter yields d_p/D_g ratios between 0.016 and 0.1. Application of this lower ratio to $D_g = 300$ to 500 μm results in a lower diameter of particles involved in bridging of 5 to 8 μm. The presence of particle bridges has been demonstrated in the field (Timmer et al. 2003; de Zwart 2007) and simulated in the laboratory (de Zwart 2007).

Besides by site specific conditions around a well, the size distribution of particles in abstracted groundwater is also influenced by filtration through the aquifer, as described by filtration theory (Logan 1999). Application of this theory for mineral particles ($\rho_s = 2600\ \text{kg/m}^3$) in an aquifer yields that particles with a diameter of 0.3 μm are most mobile, larger particles being less mobile by sedimentation and smaller particles by diffusion, and interception being negligible.

All these interactions indicate that the size distribution of particles in abstracted groundwater may only be described approximately by the empirical Pareto distribution.

According to Eq. (3.3) the slope β of the Pareto distribution may be calculated as (here for the particle size interval 2–5 μm):

$$-(\beta - 1) = \frac{\log N_{\varphi>2\mu m} - \log N_{\varphi>5\mu m}}{\log 2 - \log 5} \text{ and } \beta = 1 + \frac{\log N_{\varphi>2\mu m} - \log N_{\varphi>5\mu m}}{\log 5 - \log 2} \quad (3.4)$$

As it appears from Figure 3.6 that the particle concentrations >7 μm are underrepresented, with the help of Eq. (3.4) the slopes β have been calculated over the 2 to 7 μm interval. The results are presented in Figure 3.7. From this figure it is clear that the slope varies between 3 and 4, and is dependent on the discharge rate: the greater the discharge rate the smaller the slope. This is in line with our previous finding: an increase in flow rate results in an increase of the particle concentration, see Figure 3.2, and in a more than proportional increase of larger particles, see Figure 3.4.

Figure 3.8 shows the cumulative frequency distribution of the slopes β. In order to be sure that these values are not influenced by the under representation of larger particles, these slopes have been calculated over the 2 to 5 μm interval. This figure shows that 85% of the slopes vary between 2.65 and 3.75, with a median value of 3.1.

In literature results of other studies are mentioned: Lerman (1979) found $\beta = 4.01 \pm 0.28$ for particles in oceans, McCarthy and Degueldre (1993) $\beta = 4.2 \pm 0.2$ and Marre (2004) $\beta \approx 3.5$ for particles in groundwater, Degueldre et al. (2000) $\beta \approx 2.5$ to 6 for particles in groundwater depending on pH and dissolved organic carbon, and Borkovec et al. (1993) $\beta = 3.8 \pm 0.1$ for soil grains. Buffle and Leppard (1995) summarized data about the relationship between concentration and diameter for particles present in oceans, groundwater, lake and river water, and arrived at values close to $\beta = 3$.

Compared to these values, the order of magnitude of our values is comparable, but on the lower side. This is understandable as our results were obtained from flowing water, and most of the results in literature from stagnant water.

3.3.4 Comparison with data from monitor wells

Particles have been counted in groundwater not only from abstraction wells, but also from monitor wells. Figure 3.9 shows the particle concentration in groundwater that was slowly abstracted (100ml/min) from a monitor well (PVC, screen length 1.2 m, Ø 25 mm). This monitor well is located in a line of abstraction wells. To prevent contamination by precipitates formed in situ (such as iron hydroxides) or animal remains from isopods, beetles etc., the monitor well was cleaned and tightly closed about one month before the measurements.

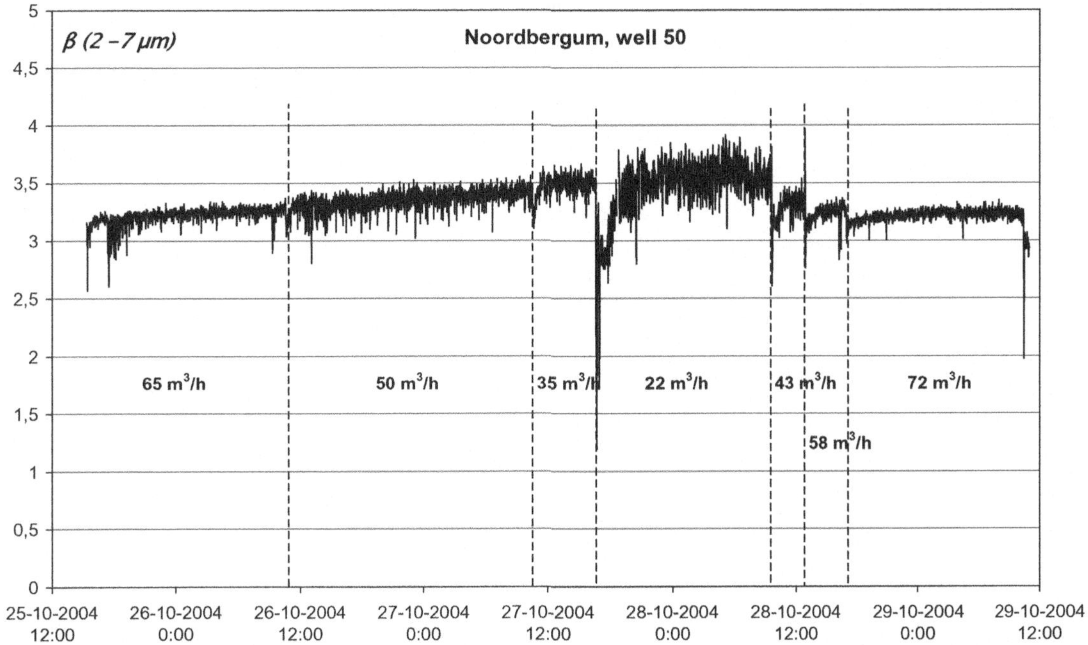

Figure 3.7 Slope β over the particle size interval 2–7 μm during continuous abstraction with variable discharge rate from well 50 of well field Noordbergum (water utility Vitens Friesland). Compare with Figure 3.2 and Figure 3.4

Note in Figure 3.9 that the concentration pattern of particles >15 μm deviates from the pattern of other diameter classes. This deviating behavior is attributed to the presence of gas bubbles, whose liquid-gas refractions are counted as particles by the particle counter. For that reason, Figure 3.9 shows concentrations of diameter intervals rather than cumulative concentrations. Gas bubbles play a prominent role here, since ground water from monitor wells is obtained by suction, while ground water from abstraction wells is obtained by positive displacement using a submersible pump.

Especially notable are the immediate and large reactions in particle concentration when the surrounding production wells are switched on. Marre (2004) made the same observation when an abstraction well near his monitor well was switched on. Since there are no physical connections between the monitor well and the abstraction wells, for instance by means of pipe connections, this reaction cannot be caused by vibrations transmitted in these lines. The reaction must be transferred through the aquifer itself.

This immediate and strong effect on particle concentration in the surrounding groundwater can only be explained as a response to the acceleration imposed on ground water by switching on the neighboring abstraction wells. The extent of the reaction is remarkable, since the particle concentration of the diameter interval 2–3 μm before 8 AM was only about 11/ml, and after switching on the neighboring wells went up to 4000/ml. It takes a long time before the concentration returns to its background value.

Due to the distance of production wells 2 and 6 to the monitor well, the effect of switching these wells on and off on the ground water flow velocity is small ($q = Q/(\pi\,\varepsilon\,dL) = 80/(\pi * 0.3 * 100 * 24) \approx 0.03$ m/h). Switching off abstraction well 4 also has a notable effect on the particle concentration: the concentrations of small particles decrease, but those of large particles seem to increase. For the moment we ascribe this phenomenon to the increased formation of gas bubbles.

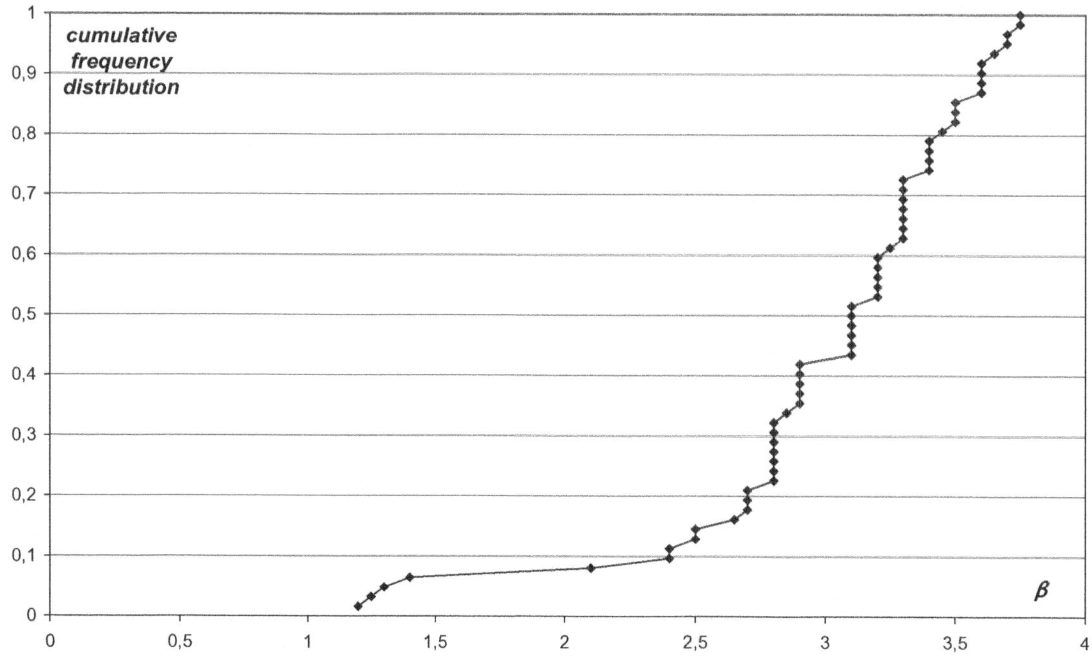

Figure 3.8 Cumulative frequency distribution of the "Pareto" slopes β calculated over the particle size interval 2 to 5 μm, in abstracted groundwater under normal operating conditions

It is also notable that the particle concentration in the diameter interval 2 to 15 μm in the monitor well before the experiments amounts to around 20/ml, and the particle concentration with a diameter >2 μm in abstraction well to 4 to 10/ml (de Zwart 2007). Apparently the particle concentration in vertical direction over the height of the aquifer is not constant (de Zwart 2007).

Hofmann (1998), Marre (2004) and Dehnert et al. (2003) also measured particle concentrations in groundwater obtained from monitor wells. However their measurements were performed by batch particle counts instead of online counts (Hargesheimer and Lewis 1995). At regular time intervals they took samples, which were subsequently counted, yielding the particle concentration and a very detailed particle size distribution.

Hofmann (1998) conducted his experiments on three monitor wells located on a well field. Under these conditions he did not detect any particles >2 μm in groundwater, which he ascribed to exhaustion of the particle supply by groundwater abstraction during 40 years. Moreover he found fewer particles in aerobic groundwater than in anaerobic groundwater. This last finding was corroborated by the results of Marre (2004) and Ryan and Gschwend (1990).

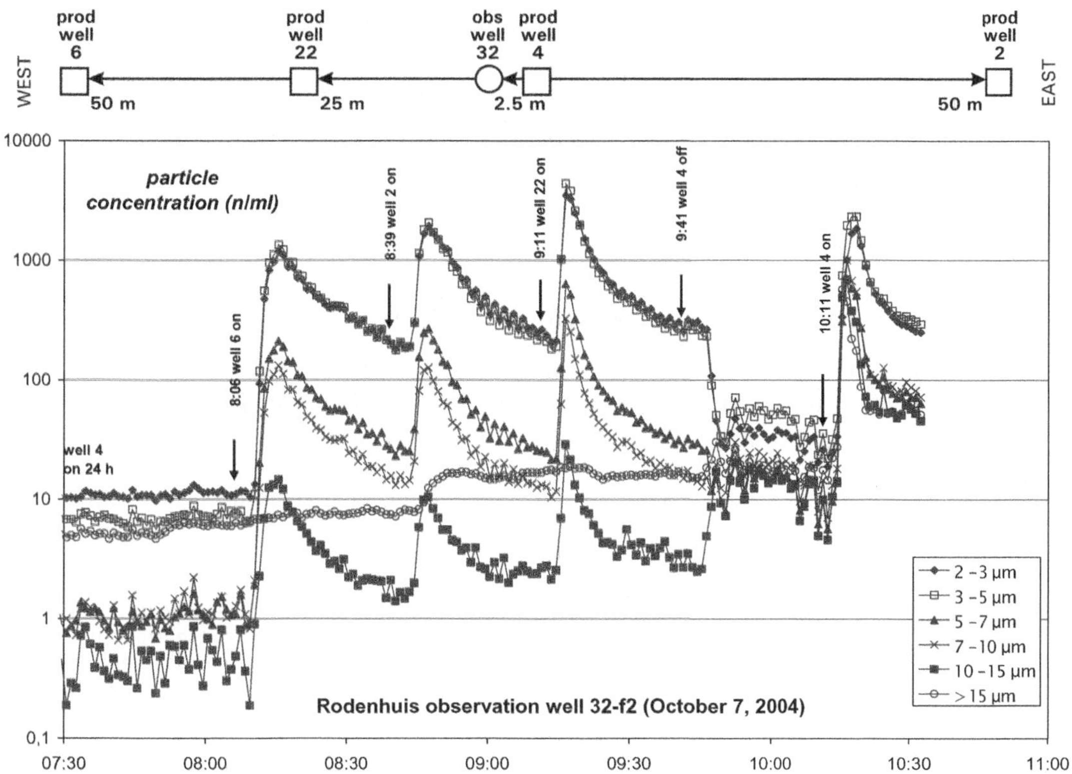

Figure 3.9 Particle concentrations in groundwater from a monitor well as a result of switching neighboring abstraction wells on and off. All wells are located in one line; the distance between the wells is shown in the inset. The delay in reaction of monitor well 32 when switching the abstraction wells on and off is caused by the time needed to displace the volume of water in the monitor well and in the sample tube. The screen of abstraction well 4 extends from 14.4 to 38.4 mbs, and of monitor well 32 from 34.4 to 35.6 mbs. Note that the particle concentration represents diameter intervals rather than cumulative particle diameters

In his research, Marre counted particles with a diameter >2.58 µm. Substitution of $\varphi > 2$ µm and $\varphi > 2.58$ µm in (3), subtraction and some reworking yields:

$$N_{\phi>2\mu m} = N_{\phi>2.58\mu m}\left(\frac{2.58}{2}\right)^{(\beta-1)} \tag{3.5}$$

Recalculating his particle concentrations with Eq. (3.5) and $\beta = 3.5$ to particle concentrations with a diameter >2 µm yields values of several thousand particles/ml for anaerobic ground waters.

Dehnert et al. (2003) abstracted groundwater from monitor wells with various discharge rates (0.2–1.2 m^3/h). They measured the concentration of particles with a diameter >0.5 µm and arrived at particle concentrations between $0.25 * 10^6$ and $2 * 10^6$/ml. Recalculating their concentrations with $\beta = 3.5$ to particle concentrations with a diameter >2 µm yields values of 10,000 to 50,000/ml.

McCarthy and Degueldre (1993) presented a summary of data on the presence of particles in groundwater, including some from unconsolidated alluvial deposits. They reported their results as mass concentrations, which usually ranged between <1 to 6 mg/l, but reached a maximum of 60 mg/l. Assuming a particle concentration >2 μm of 100/ml, an order of magnitude of the particle mass concentration was estimated for $\beta = 3$ and $\beta = 4$. These estimations show that, even with a large upper diameter of 50 μm, particle mass concentrations >0.45 μm for $\beta = 3$ and $\beta = 4$ correspond with 0.066 and 0.020 mg/l respectively. These values are comparable with the lowest mass concentrations summarized by McCarthy and Degueldre (1993). For particles >2 μm, these mass concentrations correspond with 0.064 and 0.014 mg/l respectively; or 1 mg/l corresponds with 1500 to 7500 particles >2 μm/ml, which is comparable with the concentrations in anaerobic ground waters reported by Marre (2004).

3.3.5 Discharge rate and particle concentration

The relation between particle concentration and discharge rate has been measured on several well fields, and is summarized in Figure 3.10. This figure shows that on average the particle concentration initially increases slowly with discharge rate, and subsequently increases steeply. An increase in discharge rate corresponds with an increase in pore flow velocity, resulting in higher concentrations (see Figure 3.2) especially of larger particles (see Figure 3.4). A decrease in discharge rate results in an opposite reaction.

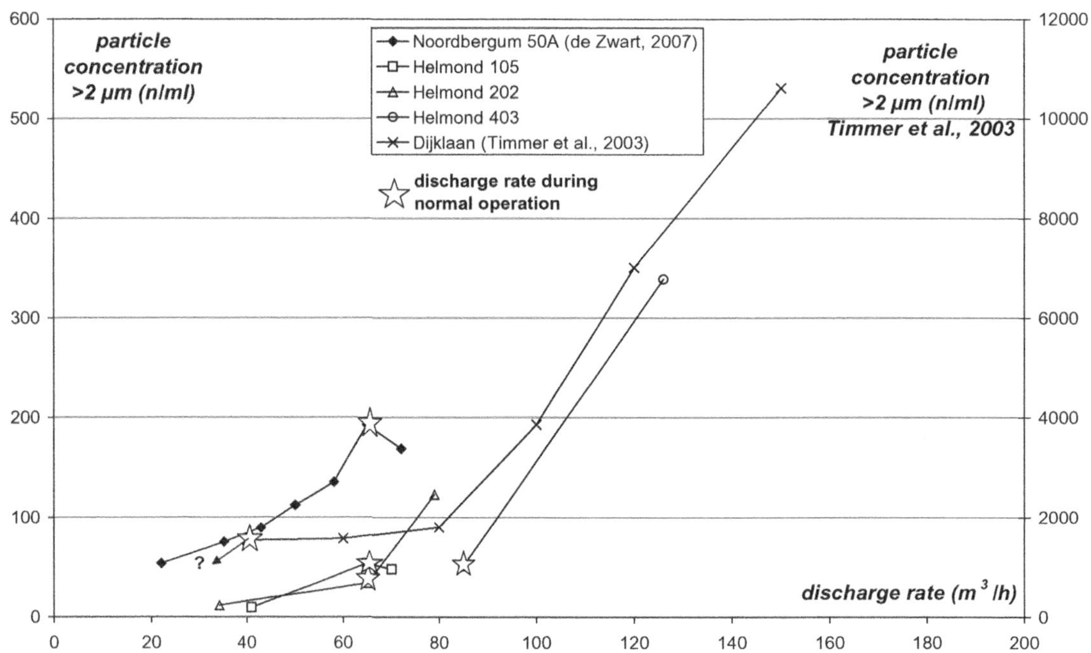

Figure 3.10 Relationship between particle concentration(> 2 μm) in abstracted groundwater and discharge rate

Moreover, a distinction should be made between discharge rates that are higher or lower than usual. In order to set a particle in motion, a critical pore flow velocity should be exceeded. Increasing the discharge rate beyond the usual rate extends the distance of this critical pore flow velocity further away from the well in the aquifer and taps a particle pool that is not available during usual operation. This results into a steep increase in particle concentration. Decreasing the discharge rate below usual results in an opposite reaction: the pore flow velocity remains high enough to keep smaller particles in motion, but is no longer able to move some larger particles. This results into a slow decrease in particle concentration. This effect will be enhanced by the exhaustion of the particle pool close to the well, resulting in particle concentrations approaching zero at low discharge rates. According to this reasoning, the usual discharge rate functions as a turning point. Except for the lower values of Dijklaan, the data in Figure 3.10 are reasonably compatible with this picture.

3.4 CONCLUSIONS

Groundwater abstracted from anoxic unconsolidated sandy sediments contains particles >2 μm in concentrations from 2 to 120/ml.

Continuous switching on and off the (submersible) pump goes along with continuous mobilization of particles filtrated by the aquifer soil matrix, in this way promoting their transport towards the abstracting well. Consequently, results of transport studies under stationary conditions in soil columns (Close et al. 2006; Hornberger et al. 1992) or in the field (Harvey et al. 1989; van der Wielen et al. 2008; Bales et al. 1997) underestimate the mobility of particles under these dynamic field conditions.

The particle concentration is related to the groundwater flow velocity: the higher the velocity, the higher the concentration. The usual discharge rate probably functions as a turning point: at higher discharge rates the radius of influence is extended further into the aquifer, and a particle pool is tapped that is not available during normal operation. This results in a steep relationship between particle concentration and discharge rate. Near the well the particle pool may already be more or less depleted, resulting in a more horizontal relationship. As a consequence, the particle concentration is governed by the initial amount minus the amount removed during well operation. As the particle supply around each well is unique, there will be no general relationship between particle concentration and groundwater flow velocity or discharge rate. Moreover, the condition of the well may influence this relationship. For example, well clogging could lead to locally higher groundwater flow velocities.

The particle size fraction >3 μm varies between 0.45 and 0.67, with a median value of 0.57. The median values for the fractions >5 μm and >10 μm equal respectively 0.14 and 0.02. The particle size distribution too is related to the groundwater flow velocity: the higher the velocity, the greater the proportion of larger particles.

The particle size distribution may be described by the power law or Pareto distribution. In our research, all Pareto distributions show an under-representation of particles with a diameter >7 μm. This break in slope may be caused by mixing of particle size distributions from finer and coarser sandy layers in the aquifer, and/or by the formation of particle bridges, in particular near the gravel pack/aquifer interface.

The Pareto slope β over the particle size interval from 2 to 5 μm varies between 2.65 and 3.75, with a median value of 3.1. Compared with values from literature, β up to about 4, our value is on the lower side. Contrary to most values in literature, our results were obtained from flowing water: the greater the flow velocity, the greater the proportion of larger particles, and the lower the value of slope β.

Except for one site (Hofmann 1998), groundwater obtained from monitor wells with minimal disturbance during sampling contained more particles >2 μm than groundwater obtained from abstraction wells during operation: 2,500 to 10,000/ml (Dehnert 2003; Marre 2004) vs 2 to 120/ml. This contradicts our conclusion that the higher the flow velocity, the higher the concentration. This discrepancy can only be explained by the exhaustion of the particle pool around the well in the aquifer by groundwater abstraction, as suggested previously by Hofmann (1998).

Chapter 4

Mineralogy and generation of particles in anoxic groundwater abstracted from unconsolidated sedimentary aquifers

C.G.E.M. van Beek[1], C.G. van Beek[2] and P.J. Stuyfzand[1,3]

[1]*KWR Watercycle Research Institute (formerly: Kiwa Water Research), P.O. Box 1072, 3430 BB Nieuwegein, the Netherlands; +31 30 60 69 511; fax +31 30 60 61 165; Kees.van.Beek@kwrwater.nl; Pieter.Stuyfzand@kwrwater.nl*

[2]*ASPEX Corporation, Delmont, PA; kvanbeek@aspexcorp.com*

[3]*VU University, Amsterdam, pieter.stuyfzand@falw.vu.nl*

ABSTRACT

More than 6000 particles >2 μm, filtered from anoxic (sulfate reducing) groundwater from unconsolidated sedimentary aquifers from 10 wells at 5 sites were analyzed by automated Scanning Electron Microscopy-Energy Dispersive X-ray Spectroscopy (SEM-EDS). This method enables detailed objective morphological and elemental characterization of these particles. The mineralogy was deduced from the elemental composition; minerals recognized include quartz, cation silicates, kaolinite, cation alumino silicates like feldspars, micas and clay minerals, calcite and iron sulfide. Calcite and silicates were present in amounts varying from 1 to 92% calcite and from 7 to 79% silicates. Moreover, metal sulfide particles as corrosion products from the well construction and/or sample valve, and metal particles as erosion (wear) products from the submersible pump were identified. No manganese oxide or iron (hydr)oxide particles were observed. As these particles are unstable under iron-reducing conditions, these observations show that mineral particle distribution is governed by aquifer mineralogy and groundwater chemistry.

Compared to the aquifer grain matrix, calcite particles are over-represented with respect to quartz. Due to redox processes producing alkalinity, groundwater already saturated with respect to calcite will become super saturated, and generate calcite particles. As soon as no alkalinity is produced anymore, this production mechanism stops. As calcite particles cannot be generated chemically anymore, but the presence of these particles remains dominant, they must be generated in another, mechanical, way, i.e. by crushing and abrasion of grains by the weight of the overlying soil column. This reasoning explains the overrepresentation of calcite over quartz particles: quartz is much harder than calcite and consequently will be more resistant to fragmentation, while calcite particles are also generated by chemical precipitation.

Information on particle mineralogy is important for estimating effects of particle facilitated transport of pollutants, bias in groundwater sampling, and clogging of wells and aquifers.

4.1 INTRODUCTION

Suspended solids in groundwater are considered as possible carriers of pollutants (McDowell-Boyer et al. 1986; McCarthy and Zachara 1989; Ouyang et al. 1996) like hydrophobic organic compounds (Enfield and Bengtson 1988; Hofmann 1998; Roy and Dzombak 2000), pesticides (de Jonge et al. 1998), metals (Mills et al. 1991) and radio-active elements (Saiers and Hornberger 1996), and as potential bias in groundwater sampling (Stuyfzand 1987; Puls and Powell 1992; Backhus et al. 1993). Information on behavior of suspended solids may also be important for understanding transport of (pathogenic) micro organisms in soils (Bales et al. 1997). Particles are also responsible for clogging of groundwater abstraction wells by accumulation on the well bore (Chapter 5), of groundwater recharge wells (Huisman and Olsthoorn 1983) and of aquifers (Vaidya and Fogler 1990).

Mineral particles in groundwater from sedimentary aquifers consist of quartz, silicates, clay minerals like kaolinite, micas like muscovite, feldspars, titanium oxide, calcite, iron hydroxides like goethite, manganese oxides, aluminum hydroxides, in situ precipitates like iron phosphates, rock fragments and other weathering products (McCarthy and Degueldre 1993; Ryan and Elimelech 1996; Seaman et al. 1997; Degueldre et al. 2000; Goldberg et al. 2000; McCarthy and McKay 2004). In field and laboratory column experiments with ferric oxyhydroxide-coated aquifer sediment of Cape Cod (USA), Bunn et al. (2002) mobilized colloids consisting of quartz, feldspar and kaolinite, together with small amounts of smectite and illite/muscovite. Goethite and other ferric oxyhydroxides were not detected.

The objective of this study is to estimate the mineral distribution of particles present in abstracted anoxic (sulfate reducing) groundwater from unconsolidated sedimentary aquifers. This information provides better understanding of the transport and clogging properties of these particles. The second objective of this study is to provide information about their generation in aquifers.

In order to arrive at this information, the elemental composition and morphology of 12177 particles were analyzed, of which 6063 particles >2 μm were evaluated, with the help of automated SEM-EDS, Scanning Electron Microscopy coupled with Energy Dispersive X-ray Spectroscopy (Schamber and van Beek 2006). This setup and equipment allow detection and quantification of elements carbon and heavier. The automated procedure makes it possible to analyze large numbers of particles from one sample in a consistent, objective and time effective way. As far as known this method has not been applied previously for determining the mineral distribution of particles in (ground)water.

Atteia et al. (1998) applied a comparable method for the identification of particles in surface water of two small brooks. However their method was not automated, and their apparatus was equipped with a beryllium window, limiting detection and quantification to sodium and heavier. After carbon coating of the polycarbonate filters, they analyzed in total approximately 1000 randomly selected particles. Peng et al. (2004) applied the same automated method for the identification of particles in surface water intended for New York city drinking water. Chanudet and Filella (2006b, 2008) applied Transmission Electron Microscopy (TEM), EDS and Selected Area Electron Diffraction (SAED) for the identification of mineral particles in surface water and in drainage water (Filella et al. 2009). They analyzed 100 particles per sample; analysis of more samples would lead to more cost and also possibly to operator fatigue (Chanudet and Filella 2006a). Furthermore, TEM requires samples to be prepared as thin, typically 100 nm, films. In other studies, electron microscopy has been applied for the identification of a limited number of particles, too few to calculate the mineral distribution.

4.2 FIELD SITES AND METHODS

4.2.1. Site description

At each of the 5 well fields studied (Figure 4.1, Table 4.1 for identification), two wells were sampled. The selected well fields were involved in earlier research on mechanical clogging of groundwater abstraction wells, incl. development and rehabilitation (van Beek 1984; Breedveld et al. 2007), micro morphology of soil samples of the well bore and adjacent aquifer (Timmer et al. 2003; de Zwart 2007) and particle counting (see Chapter 3). The sampled wells were in operation for a few to almost 40 years (Table 4.1) for the public drinking water supply. Over the years, disturbances caused by well drilling must have faded away. Consequently, the presented results are representative for groundwater abstraction conditions.

Table 4.1 shows a summary of some construction data of the wells involved. Remarkable is the screen of SO21 (=well field Someren well number 21), with a total screen length of 28.4 m divided in 7 parts over 98 m.

Table 4.2 shows the geology of the aquifers involved, varying from ancient leached fluvial sediments, poor in reactive material like organic matter and calcite, to recent fluvial deposits, rich in reactive material. Note that the aquifers at NB and SO contain no calcite. All aquifers contain iron sulfides in variable amount. Huisman (1998) mentions the presence of glauconite in the Tertiary deposits, but this presence was not confirmed by the drilling logs. He also mentions the presence of iron carbonate, but not of iron (hydr)oxides. The aquifer depth is indicative for the age of the abstracted ground water and has been indicated in Table 4.2. Groundwater is abstracted below extensive clay layers, except for RH, abstracting riverbank filtrate, and NB52A, with nearby a weak spot in the confining clay layer.

Table 4.3 presents the chemical composition of the abstracted groundwater close to the time of sampling, i.e. June 2006. However, except for RH04, RH06 and NB52A, the chemical composition of the abstracted groundwater does not change over time. As evidenced by $SI_{calcite}$, all waters except SO are practically calcite saturated. Moreover all waters are strongly supersaturated with respect to quartz (solubility at 25°C 0.1 mmol/l = 6 mg/l as SiO_2; Stumm and Morgan 1996). Sulfate concentrations < 0.05 mmol/l indicate nearly complete sulfate reduction, and higher concentrations as in NB52A and RH to partial reduction. All waters contain methane, indicating that redox conditions must be, at least locally, methane producing. The temperature varies between 10°C for shallow (RH) to 13°C for deep wells (SO).

4.2.2 Sampling

The sampled groundwater passed subsequently the aquifer, well bore, gravel pack, well screen, well pipe, submersible pump, riser pipe, sampling valve inclusive connection, and sampling tube (Appendix Figure A4.1). All these parts may give off particles. In order to obtain representative samples (see Chapter 3), the wells were operated several hours prior to sampling and the sampling valve flushed. During sampling, water from this valve was led under pressure (of the submersible pump) over a 0.45 μm polycarbonate membrane filter. In order to obtain separate particles, and to prevent stacking of particles (Seaman 2000), the sampling time was estimated from prior particle counts. Taking arbitrarily 1% as the maximal covering of the membrane filter by particles, calculated sampling times varied from 20 to 40 minutes, equivalent to 2 to 5 l water. After sampling, the (wet) filter was transferred to a Petri dish and air dried.

4.2.3 Analysis

Samples were analyzed with automated Scanning Electron Microscopy/Energy Dispersive X-ray Spectroscopy (SEM/EDS). In SEM a focused beam of electrons is rastered over the sample, resulting in

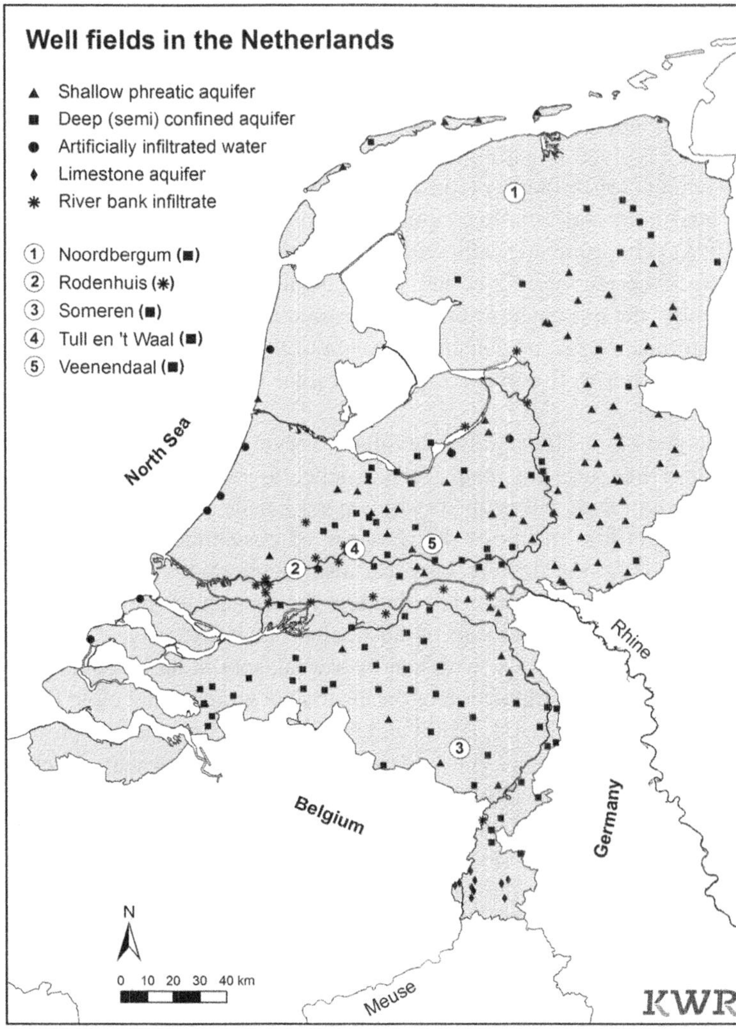

Figure 4.1 Map showing the well fields operated by the Dutch waterworks, and the well fields under consideration

emission of various signals. Of these emitted signals, the fraction of backscattered electrons is proportional to the average atomic number at the site of the impinging electron beam. EDS displays characteristic peaks for each element corresponding to that element's unique electron transitions. The peaks from all elements in the interaction volume of the impinging electron beam and sample are simultaneously acquired. Once a particle has been identified based on backscattered electron contrast, the beam looks for the centre of this particle. Starting from this centre, the particle is analyzed proportionally distributed (over 360^0) 16 times, yielding morphological and chemical characteristics. The method has been described in more detail by Schamber and van Beek (2006).

The morphology was recorded in accordance with the "Standard practice for characterization of particles" (ASTM 2003), and consisted of average diameter (d_{ave}), maximum diameter (d_{max}), minimum diameter (d_{min}), perpendicular diameter being the particle diameter perpendicular to the maximum diameter (d_{perp}), area of the particle (A), and perimeter of the particle (p).

Table 4.1 Construction data of the 10 investigated abstraction wells. (mbs: meter below surface)

Well field (Site)	Well code	Year of construction	Drilling Method[&]	Well screen			
				Depth (mbs[@])	Total length (m)	Number of segments	Material
Noordbergum	NB50B*	2004	rral	67.5–86.7	19.2	1	PVC
(NB)	NB52A	1992	rral	50.5–83.3	30.8	1	PVC
Rodenhuis	RH04	1967	b	17.3–40.9	23.6	1	SS[#]
(RH)	RH06	1967	b	17.4–39.1	21.7	1	SS
Someren	SO21	1998	rral	163.0–261.0	28.4	7	PVC
(SO)	SO22	1998	rral	193.4–261.8	29.1	4	PVC
Tull en 't Waal	TW22	1972	rral	129.0–144.0	15.0	1	SS
(TW)	TW33	2004	rral	84.4–121.0	22.6	3	PVC
Veenendaal	VE14	1977	rral	92.0–107.0	15.0	1	SS
(VE)	VE22	2004	b	96.5–117.0	17.5	2	wws**

*NB50B: well field Noordbergum well number 50B;
[&]b: bailer; rral: reverse rotary – air lift, with minimal drilling mud
[@]mbs: meters below surface
[#]SS: (asphalt coated) stainless steel
**wws: wire wrapped screen

Depending on the thickness of the particle and the density of the composing atoms, the interaction volume (depth) of the impinging electrons may exceed this thickness. This phenomenon is also relevant as soon as the beam reaches the rim of the particle. Consequently the EDS obtained includes signals from the polycarbonate membrane filter below and around the particle: the smaller (the thinner) the particle, the larger this effect. As mineral particles may also contain C and O, and as it is not possible to distinguish between both origins, this phenomenon complicates mineral identification.

To allow analysis of nonconductive samples, the uncoated samples were placed inside the ASPEX® PSEM™ and a sample pressure of 0.15 Torr was maintained around the samples throughout the analysis. This low pressure gaseous environment allows the current from the impinging electron beam to be conducted away from the sample instead of accumulating on the sample. This allows additional impinging electrons to reach the sample at the required location. Alternative techniques to dissipate the electron charge are carbon or Au/Pb coating. The selected experimental setup allowed for 1500 particles to be characterized in 2.5 hours. To allow for the distribution of the particles over the full membrane filter surface, and hence time to scan the empty membrane between particles, each sample was analyzed for 1500 particles with a maximum of 5 hours, whichever criterion was reached first. During this time the sample was scanned with an electron beam of 25 kV accelerating voltage and 0.5 nA sample current. A working distance of 16.8 mm was maintained throughout the analysis. Each sample was scanned in multiple 178 μm square fields using a step size of 0.35 μm. Only features with a signal higher than the set threshold were analyzed and their energy dispersive spectra acquired for a maximum of 6 seconds. The acquired spectra were compared with standard reference spectra containing the following elements: Al, As, Ba, Bi, C, Ca, Cl, Co, Cr, Cu, Fe, K, Mg, Mn, Na, Ni, O, P, Pb, S, Si, Sn, Sr, Ti, V and Zn.

Table 4.2 Some (geological) characteristics of the well fields

>Well field Residence time	Geological Formation	Geological description	$CaCO_3$ wt%
Noordbergum (NB) $\tau^{\#}$ = ca 100 y	Enschede interfingering with	Leached coarse sandy river Rhine deposits, containing gravel	<0.1
	Harderwijk (middle Pleistocene)	Leached coarse sandy river Elbe and Weser deposits	<0.1
Rodenhuis (RH) $\tau = 3 - 10$ y	Kreftenheije on Sterksel (late Pleistocene)	Recent river Rhine deposits, rich in weatherable minerals and organic matter	1–2 1–2
Someren (SO) τ = ca 10.000 y	180–200 mbs Kiezeloöliet, Peij sands (late Tertiary)	Leached deposits with intercalated clay and brown coal layers	<0.1
	225–250 mbs Kiezeloöliet, Waubach sands (late Tertiary, Miocene)		<0.1
Tull en't Waal (TW) τ = 2000 à 3000 y	Harderwijk interfingering with Tegelen (middle Pleistocene)	Leached river Elbe and Weser deposits	<0.1
		River Rhine deposits containing weatherable minerals and organic matter.	1–3
Veenendaal (VE) τ = 2000 à 3000 y	Tegelen (middle Pleistocene)	River Rhine deposits containing weatherable minerals and organic matter.	1–3
	Maassluis (early Pleistocene)	Fine sandy marine deposits, containing shells	3–5

*mbs: meter below surface
#τ: indicative age of the abstracted groundwater

By dividing the area under the characteristic EDS peak of an element by the total area under all signals, a relative content of the elements present is obtained: the atomic (or molar) weight percentage (at%). To account for "Bremsstrahlung" a filter-fit was applied (McCarthy and Schamber 1981). While the chemical parameters were selected to provide a balance between obtaining statistically representative results of the particle population and the chemical composition of the individual particles, experience showed that concentrations <1 at% are not statistically significant. Thus, if the relative area amounts to <1%, the contribution of the element involved is ignored, and the distribution calculated again. As the EDS also contains signals from the polycarbonate membrane, this 1% cut off value becomes larger with respect to mineral content. In particular for small particles, this lower limit may increase to 4–5%.

4.2.4 Identification and distribution of mineral particles

The EDS analysis discussed above yields information about which elements are present in a particle, together with an approximate atomic content (heavy elements contain more electrons, resulting in more radiation and consequently in a larger area under the curve, and in this way in an overestimation compared to light elements). In this way an approximate elemental composition of each particle is obtained. By comparing the elemental composition of the particles with those of minerals, it is possible to assign each particle to a mineral. In this assignment general information about the presence of minerals in Dutch aquifers was used (Breeuwsma 1990; Huisman 1998). However, no mineralogical methods were applied to confirm this assignment. Here, for convenience, each mineral will be named after its crystalline state: quartz for silicon dioxide and calcite for calcium carbonate.

Table 4.3 Chemical composition of abstracted groundwater from the denoted wells

para-meter date	Unit	NB50B Apr-05	NB52A Apr-05	RH04 Sep-05	RH06 Sep-05	SO21 Oct-05	SO22 Nov-06	TW22 Oct-03	TW33 Jun-05	VE14 Oct-05	VE22 Oct-05
EC	mS/m	56	57	62.0	65.1	31.0	33.6	40	37	38	29
pH		7.15	7.30	7.40	7.45	6.76	6.70	7.7	7.7	8.0	8.0
Cl	mmol/l	1.28	1.51	2.15	2.71	0.25	0.25	0.34	0.25	1.13	0.73
NO_3	mmol/l	<0.016	<0.016	<0.002	<0.002	n.a.	n.a.	<0.002	<0.002	<0.002	<0.002
SO_4	mmol/l	<0.10	0.33	0.37	0.48	0.03	<0.04	<0.01	<0.01	0.04	0.05
HCO_3	mmol/l	5.39	4.26	3.79	3.43	3.46	3.75	4.31	4.15	2.89	2.71
CO_2	mmol/l	1.02	0.57	0.39	0.32	1.36	1.59	0.25	0.21	0.09	0.07
Fe	mmol/l	0.134	0.130	0.046	0.039	0.091	0.116	0.015	0.015	0.006	0.005
Mn	mmol/l	0.004	0.005	0.010	0.010	0.004	0.004	0.002	0.003	0.001	0.001
NH_4	mmol/l	0.043	0.014	0.071	0.068	0.033	0.022	0.014	0.009	0.022	0.020
SiO_2	mmol/l	0.22	0.20	0.20*	0.20*	0.35*	0.35*	0.33*	0.31	0.30*	0.30*
PO_4-ortho	mmol/l	0.003	0.002	0.009	0.007	0.000	0.000	0.001	0.001	0.001	na
TOC	mmol/l	0.44**	0.42	0.21	0.18	0.23	0.17	0.18#	0.17#	0.07#	0.07#
Na	mmol/l	0.96	1.00	2.03	2.41	0.48	0.61	0.61	0.44	1.30	1.09
K	mmol/l	0.06	0.05	0.10	0.09	0.04	0.07	0.03	0.02	0.17	0.14
Ca	mmol/l	2.30	2.18	1.91	1.71	1.23	1.18	1.75	1.78	0.93	0.85
Mg	mmol/l	0.36	0.30	0.45	0.40	0.23	0.35	0.24	0.24	0.46	0.32
TH$	mmol/l	2.66	2.46	2.35	2.10	1.44	1.52	1.99	1.99	1.35	1.17
Al	μmol/l	0.15*	0.15*	0.07*	0.07*	0.07*	0.07*	<0.04*	<0.04*	<0.04*	<0.04*
$SI_{calcite}$		−0.09	−0.06	−0.07	−0.13	−0.75	−0.80	0.25	0.26	0.10	0.06
CH_4	μmol/l	369	37	24	16	94	34	111	158	0.6	na
Res Alk&	$mmol_c$/l	0.07	−0.66	−0.91	−0.77	0.58	0.71	0.33	0.17	0.19	0.37

*average all wells, Jan 2008; **April 2006; #: DOC; n.a.: not analyzed; $Total Hardness, Ca + Mg; &Residual Alkalinity, HCO_3 – 2TH.

Dutch aquifers consist for 80 to 95% of quartz. Other silicate minerals, which may be present, with average chemical composition (Gaines et al. 1997), include feldspars like potassium feldspar: KSi_3AlO_8, and sodium feldspar: $NaSi_3AlO_8$, muscovite: $KAl_2(Si_3AlO_{10})(OH)_2$, biotite: $K(Fe^{2+},Mg)_3(Si_3AlO_{10})(OH,F)_2$, augite: $(Ca,Mg,Fe^{2+},Fe^{3+},Al,Ti)_2(Si,Al)_2O_6$, hornblende: $Ca_2(Mg,Fe^{2+})_4Al\ (AlSi_7O_{22})(OH)_2$, glauconite: $(K,Na)(Mg,Fe^{2+})_{0.33}(Fe^{3+},Al)_{1.67}(Si,Al)_4O_{10}(OH)_2$, and clay minerals like illite: $K(H_3O^+)Al_2(Si_3AlO_{10})(OH)_2$ and kaolinite: $Al_2Si_2O_5(OH)_4$.

Carbonates may be present as calcite: $CaCO_3$, dolomite: $CaMg(CO_3)_2$ or siderite: $FeCO_3$. Calcite is mainly present as a secondary mineral as a concretion or as a coating on mineral grains or as shell debris, and hardly as a primary mineral. Hydroxides may be present as iron-hydroxides, predominantly as secondary minerals as a coating. Sulfides may be present as iron sulfides (pyrite: FeS_2, or mackinawite: FeS), usually generated as a secondary phase.

After assigning each particle to a mineral, the particle mineral frequency distribution may be calculated. For this calculation a procedure (decision tree) was developed. In this procedure minerals are classified according to their stability, starting with silicates, via carbonates and sulfides to the least stable (artifact) phase, chlorides. If two or more anions are present, the particle is assigned to the most stable phase. In this way all particles are counted only once. The procedure (decision tree) reads as follows (Figure 4.2):

1. Only silicon present (Si $>$ 1%, Al, Fe, Na, K, Ca and Mg $<$ 1%): quartz.
2. Silicon and one or more cations like Na, K, Ca, Mg and/or Fe present, aluminum absent (Si $>$ 1%, Al $<$ 1%, Fe, Na, K, Ca and/or Mg $>$ 1%): cation silicates like olivine $(Mg,Fe)_2SiO_4$.
3. Only silicon and aluminum present, cations absent (Si and Al $>$ 1%, Fe, Na, K, Ca and Mg $<$ 1%): alumino silicates like kaolinite.
4. Silicon and aluminum present together with one or more cations Na, K, Ca, Mg and/or Fe (Si $>$ 1%, Al $>$ 1%, Fe, Na, K, Ca and/or Mg $>$ 1%): cation alumino silicates like feldspars, micas (muscovite, biotite) and clay minerals (illite, montmorillonite).
5. Only calcium present (Ca $>$ 1%, Si, S and Fe $<$ 1%): calcite.

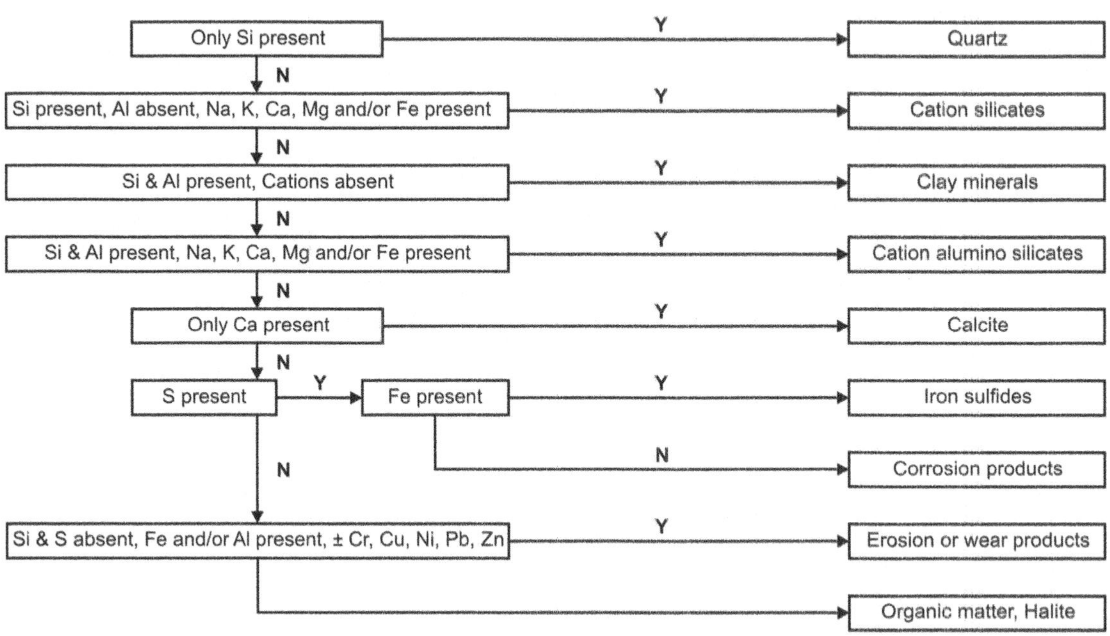

Figure 4.2 Procedure (decision tree) for determining the particle mineralogical distribution from the elemental composition

6. Sulfides.
   6a. sulfur and one or more metals present (S > 1%, Ca < 1%, Fe, Cr, Cu, Ni, Pb and/or Zn > 1%) iron sulfides and corrosion products.
   6b. iron and sulfur present (Fe and S > 1%, Ca, Cr, Cu, Ni, Pb and Zn < 1%): iron sulfides
7. Iron and/or aluminum present whether or not together with Cr, Cu, Ni, Pb and/or Zn: (Fe, Zn, Ti and/or Al >1%, Si and S < 1%): erosion or wear products. If present, iron (hydr)oxides and siderite would be assigned to this group.
8. Particles not assigned to 1–7: halite and other precipitates.

4.3 RESULTS

4.3.1 Elemental frequency distribution

Table 4.4 shows the number of particles >2 μm analyzed in abstracted groundwater from each well, and the percentage of particles containing the denoted element. The presence of carbon and oxygen has been excluded, as the results of both elements were biased by the use of a polycarbonate membrane filter. Some particles contained no elements except carbon and oxygen, and have been classified as organic matter (orgm). Appendix Table A4.1 shows the elemental frequency distribution for all particles.

Small differences in elemental frequency distribution between both wells of a well field are observed for NB and SO, whereas large differences are noted for RH, TW and VE. The large difference in TW may be attributed to abstraction from different aquifers (Table 4.1).

Table 4.4 also shows that in total 6063 particles >2 μm were analyzed, and that 51.8% of these particles contain calcium and 39.7% silicon. The extremes are represented by TW22 with 90.6% of the particles >2 μm containing calcium and SO22 with 78.5% of these particles containing silicon. NB52A represents an average with 48.6% containing calcium and 49.0% containing silicon. The percentage of mineral particles containing iron varies between 4.1 in TW22 and 48.0% in SO22, potassium between 0.6 in TW22 and 10.0% in SO22, magnesium between 1.0 in RH06 and 14.6% in RH04, sulfur between 2.5 in VE22 and 36.5% in SO21 and silicon between 6.6 in TW22 and 54.6% in SO22. Remarkable is the near absence of particles containing manganese or phosphorus in all samples (Table 4.4). Unexpected is the varying presence of various metals, like copper up to 15.1% in SO21, lead up to 9.9% in SO21, and zinc up to 16.8% in VE14 in particles >2 μm.

Surprising is the presence of chloride, between 0.2 and 51.9% of all particles, where the highest value refers to VE14. The presence of chloride is accompanied by sodium, indicating the presence of particles containing sodium chloride. This is confirmed by Figure 4.3 where for 2 sites the cation content is plotted against the chloride content, yielding a linear relationship between sodium and chloride, indicating the presence of sodium chloride containing particles. Variable Ca contents at low Cl content indicate precipitation of chloride salts on calcium containing minerals, like calcite. On the other hand, low K, Ca and Mg contents at variable Cl content indicate the presence of mixed chloride salts or the precipitation of chloride salts on mineral particles. Apparently, the presence of these salts resulted from a flaw in sampling. Drying of the wet membrane filters must have concentrated dissolved sodium and chloride to such a degree that salt precipitated. For better results, the membrane filters should have been rinsed with distilled and particle free water prior to air drying or placed shortly on a water absorbing medium. Chanudet and Fillela (2006a) centrifuged samples in the field over carbon coated, collodion covered, 200 mesh copper grids, and diluted the water samples with water where necessary. Johnson et al. (1991) were not troubled by calcium carbonate precipitation artifacts.

Table 4.4 Number of particles >2 μm analyzed, percentage of particles containing the denoted element (at % >1) and percentage of particles consisting of "organic matter"

well	>2μm	Al	As	Ba	Bi	Ca	Cl	Co	Cr	Cu	Fe	K	Mg
NB50B	757	24.0	2.5	0*	0.7	55.0	1.3	0*	0.9	0.8	21.5	5.5	4.1
NB52A	698	25.2	5.0	0.6	1.3	48.6	11.2	0*	0.7	1.3	21.9	6.1	6.4
RH04	184	28.3	2.7	1.1	0*	37.0	4.3	0*	0*	0.5	17.9	6.0	23.4
RH06	603	5.5	2.3	0.3	0.5	79.6	9.8	0*	0.2	0*	3.6	1.8	0.8
SO21	657	31.7	8.8	1.4	0.9	17.4	0.3	0*	2.6	15.1	25.1	13.2	4.3
SO22	665	43.3	10.4	2.3	1.1	4.2	0.2	0.2	2.0	1.8	29.2	16.5	1.7
TW22	1116	3.5	3.7	0.2	0.5	90.6	6.4	0*	0*	1.0	1.4	0.8	1.3
TW33	714	17.9	3.1	1.0	0.6	44.4	2.2	0*	0*	0.3	12.7	7.3	16.5
VE14	459	9.4	2.4	0*	1.3	41.8	51.9	0*	0.2	0.4	5.4	3.9	7.8
VE22	210	12.4	5.7	0*	1.0	83.3	7.1	0*	0*	2.9	10.0	6.2	19.5
total	6063	19.4	4.7	0.7	0.8	51.8	8.2	0	0.7	2.4	14.6	6.5	6.1
well	**Mn**	**Na**	**Ni**	**P**	**Pb**	**S**	**Si**	**Sn**	**Sr**	**Ti**	**V**	**Zn**	**orgm**
NB50B	0.4	2.5	1.8	2.1	0*	1.3	39.9	0.3	3.2	0.1	0*	2.0	1.2
NB52A	0.1	12.0	0.9	0.1	0*	3.3	49.0	2.6	5.6	0.4	0*	0.6	1.7
RH04	0*	9.8	0*	0.5	0*	2.7	66.3	0*	1.1	1.6	0*	2.2	3.3
RH06	0.2	10.1	0.2	0*	0*	0.3	13.6	0.2	0.5	0.5	0*	0*	0*
SO21	0*	1.1	2.3	0.5	9.9	19.2	60.7	2.0	3.2	2.0	0.2	0.2	0*
SO22	0.3	0.9	2.0	0.5	3.3	4.7	78.5	0.3	2.4	2.7	0.2	0.3	0.5
TW22	0*	6.2	0*	0*	0*	0.6	6.6	1.5	0.4	0.3	0*	0.2	0.1
TW33	0.1	2.8	0.1	0*	0*	0.8	55.0	0*	4.5	1.1	0*	3.5	1.8
VE14	0*	54.5	0*	0*	0*	17.9	20.0	0.2	0.7	0*	0*	16.8	0*
VE22	0*	11.9	0*	0.5	0*	1.4	37.6	2.9	1.9	0.5	0*	1.0	0*
total	0.1	9.2	0.8	0.4	1.4	4.9	39.7	1.0	2.5	0.9	0	2.2	0.7

*absent

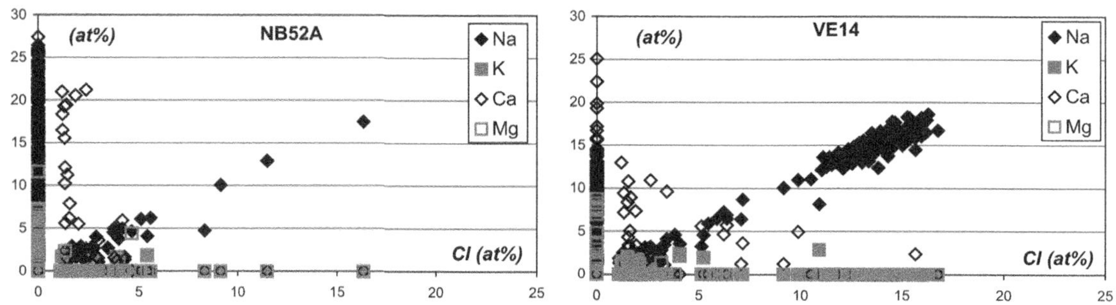

Figure 4.3 Cation content as a function of chloride content (in at%) for the indicated wells. The values on the axes near the origin are the result of assigning zero to contents >1%

Apparently, VE14 and VE22 contained most water of all samples, which resulted after drying in the high numbers of NaCl particles. If sodium chloride particles have been precipitated during drying of the sample, one may expect also other precipitates, like calcium carbonate and silica. The influence of these artifact precipitates on the results will be discussed later.

4.3.2 Particle size effects

Figure 4.4 shows the aluminum and sulfur content of particles from TW33 as a function of particle diameter. This figure shows that sulfur is predominantly present in particles <2 μm, and aluminum predominantly in particles >1 μm. Apparently, there is a size-dependent effect in the chemical composition of particles. However, looking at the aluminum concentration, this size-effect may be influenced by the cut off value of 1%.

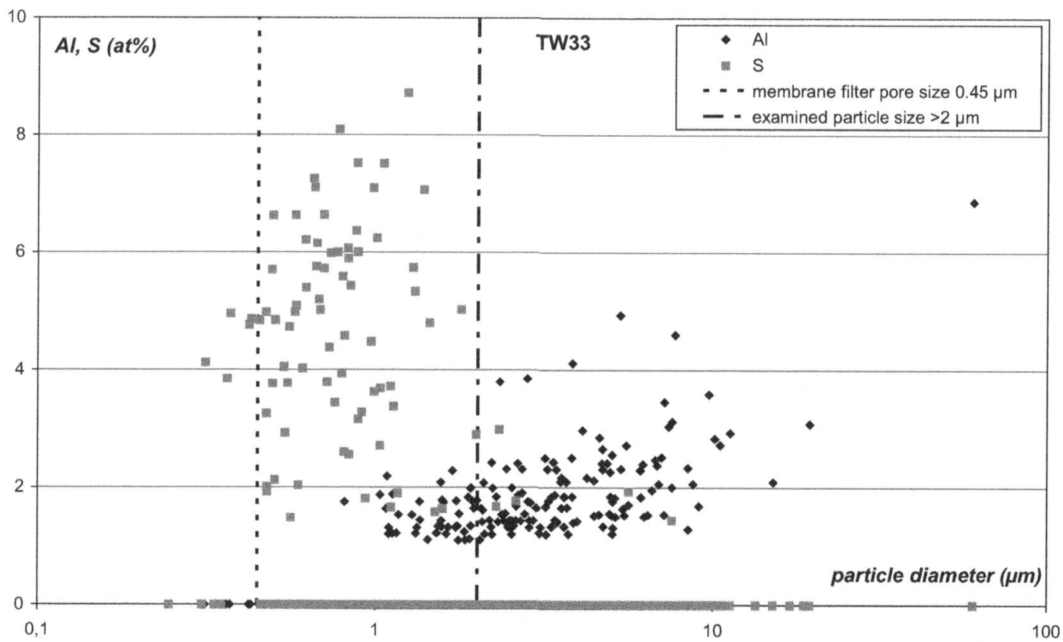

Figure 4.4 Sulfur and aluminum content of particles in abstracted groundwater from TW33 as a function of particle diameter. Observations on the horizontal axis represent diameters of particles containing no aluminum or sulfur. NB: horizontal scale is logarithmic

Figure 4.5 shows respectively the silicon and calcium content as a function of average particle diameter for the wells with the highest prevalence of silicon (SO22) and calcium (TW22). This figure clearly shows a relation between content and particle size. Consequently, the smaller the particle, the greater the contribution of the membrane filter to the EDS signal, the greater the chance that elements present in minor amounts will not exceed the 1% cut off value (see Analysis), and consequently will not be reported. In order to prevent size effects in the mutual comparison of the chemical composition of particles, only particles with an average diameter >2 μm will be considered. This diameter was also the lower boundary in previous particle counts (see Chapter 3).

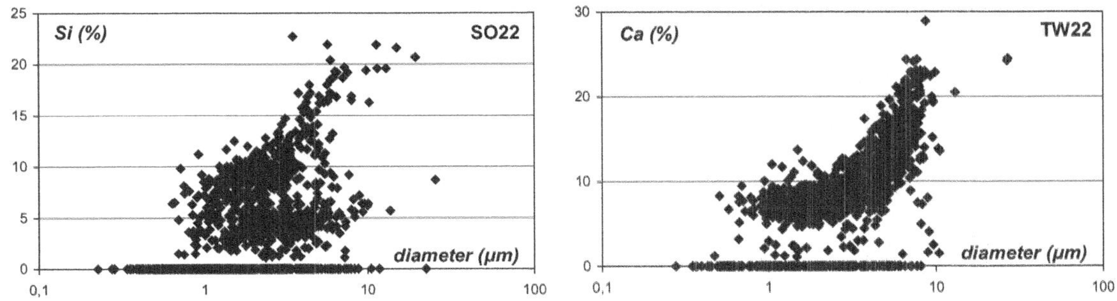

Figure 4.5 Silicon and calcium content as a function of average particle diameter in abstracted groundwater from well SO22 and well TW22 respectively. Note the presence of two populations in SO22. NB: logarithmic horizontal scale

4.3.3 Mineralogical identification of silicates

Silicates may be distinguished on the basis of their Al/Si ratio; in increasing order quartz, silicates, clay minerals, micas and feldspars. Figure 4.6 shows for NB52A, as an average site, the Al content as a function of the Si content for particles >2 µm. For other sites this relation is presented in Appendix Figure A4.2. From these figures several conclusions may be drawn.

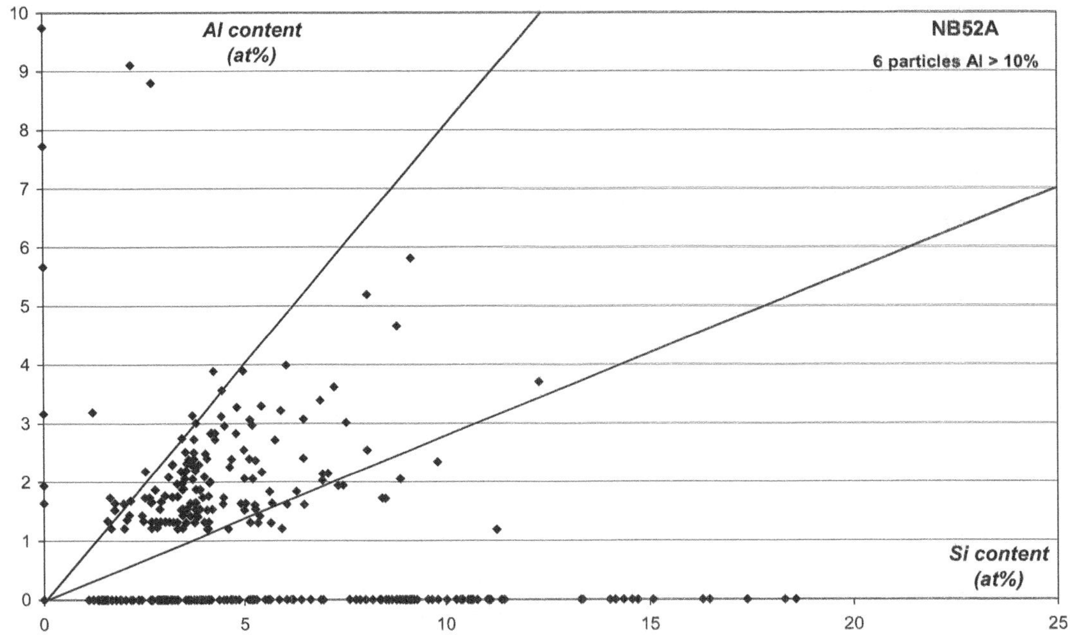

Figure 4.6 Aluminum content as a function of silicon content for particles >2 µm, in at%. Observations with Al > 10% or Si > 25% omitted (but indicated). NB: lines are drawn with naked eye

There are many observations present on the horizontal axis, indicating that many silicate particles are present without aluminum. For greater Si contents this indicates the presence as quartz or silicates. Observations near the origin may be due to assigning measured aluminum contents $<1\%$ to zero. However, due to contribution to the EDS signal of the membrane filter, the actual value may be larger, in this way somewhat overestimating the number of silicate and underestimating the number of alumino silicate particles.

There are only some observations present on the vertical axis, indicating that aluminum is almost only present as alumino silicates. Values high on the vertical axis indicate the presence of elemental aluminum. As there are hardly observations on the vertical axis near the origin, rounding down silicon contents $<1\%$ to zero plays no role.

Between the relative contents of aluminum and silicon two relationships may be distinguished, indicating the presence of roughly two different mineral groups. The slope of the upper limb is about three times as steep as the slope of the lower limb. The steep slope compares with a molar Al/Si ratio of 1, as in illite and kaolinite, and the lower slope with a molar Al/Si ratio of $\frac{1}{3}$, as in feldspars and micas. Between these two relationships, there is a gradual transition, and even some lower ratios are observed, but the maximum ratio of $Al/Si = 1$ is rarely exceeded. This gradual change in Al/Si ratio is representative for the presence of clay minerals.

There are a few observations of aluminum with low silicon. This may indicate the presence of silica precipitated on aluminum particles during drying of the membrane filter.

In order to look in greater detail to the relationship between the cations Na, K, Ca, Mg and Fe and Al and Si, these cations have been plotted as a function of the contents of aluminum and silicon. No distinction has been made between separate and combined presence, for example sodium may be present as $NaSi_3AlO_8$ or as $(Ca,Na)(Al,Si)Si_2AlO_8$. Figure 4.7 shows for NB52A the relation between Na, K, Ca, Mg and Fe and Si and Al. The relationships for all other wells are presented in Appendix Figure A4.3 to A4.7. From these figures several conclusions may be drawn.

Most notable is the difference between Na, K and Mg on the one and Ca and Fe on the other side: the near absence of Na, K and Mg on the vertical axis indicates that these cations are mainly present as (alumino) silicates, and Ca and Fe not.

Consideration of the slope of the relations between the contents of respectively Na and K, and Al and Si results in the recognition of respectively $NaSi_3AlO_8$ (albite) and KSi_3AlO_8 (potassium feldspar). For K a second relationship seems present: at equal Si and K content, the Al content is three times as high, indicating that this mineral contains three times more Al. The most probable mineral corresponding with this composition is muscovite: $KAl_2(Si_3AlO_{10})(OH)_2$. For Mg contents $>3\%$, magnesium is mainly present as a magnesium silicate. As Mg and Si seem present in nearly equimolar amounts, the mineral composition might be $MgSiO_3$ (enstatite). For contents $<3\%$ magnesium seems to be present as a magnesium alumino silicate. Based on molar relations, the chemical composition indicates biotite, K $(Fe^{2+},Mg)_3(Si_3AlO_{10})(OH,F)_2$. In TW33 magnesium appears to be present in yet another magnesium alumino silicate. Based on molar relations, magnesium is a major constituent in this mineral: $Mg:Si:Al \approx 5:2:1$.

Sometimes, K and Mg occur in low concentrations at high Si and Al contents. In these particles K and Mg are present as a minor constituent together with other cations.

Glauconite is a K- and Fe-rich and Al-poor mica of marine origin (Huisman 1998, after Deer et al. 1992). As the K-Al slope of all sites is mutually comparable and is comparable with the Na-Al slope, glauconite seems to be absent at all sites.

In some cases Na, K and Mg occur together with chloride. If these observations are on the vertical axis, they indicate the presence of pure chloride precipitates, otherwise they indicate the presence of chloride precipitates on (alumino)silicates. The observations at rather high Al concentrations probably indicate chloride precipitates on elemental aluminum particles.

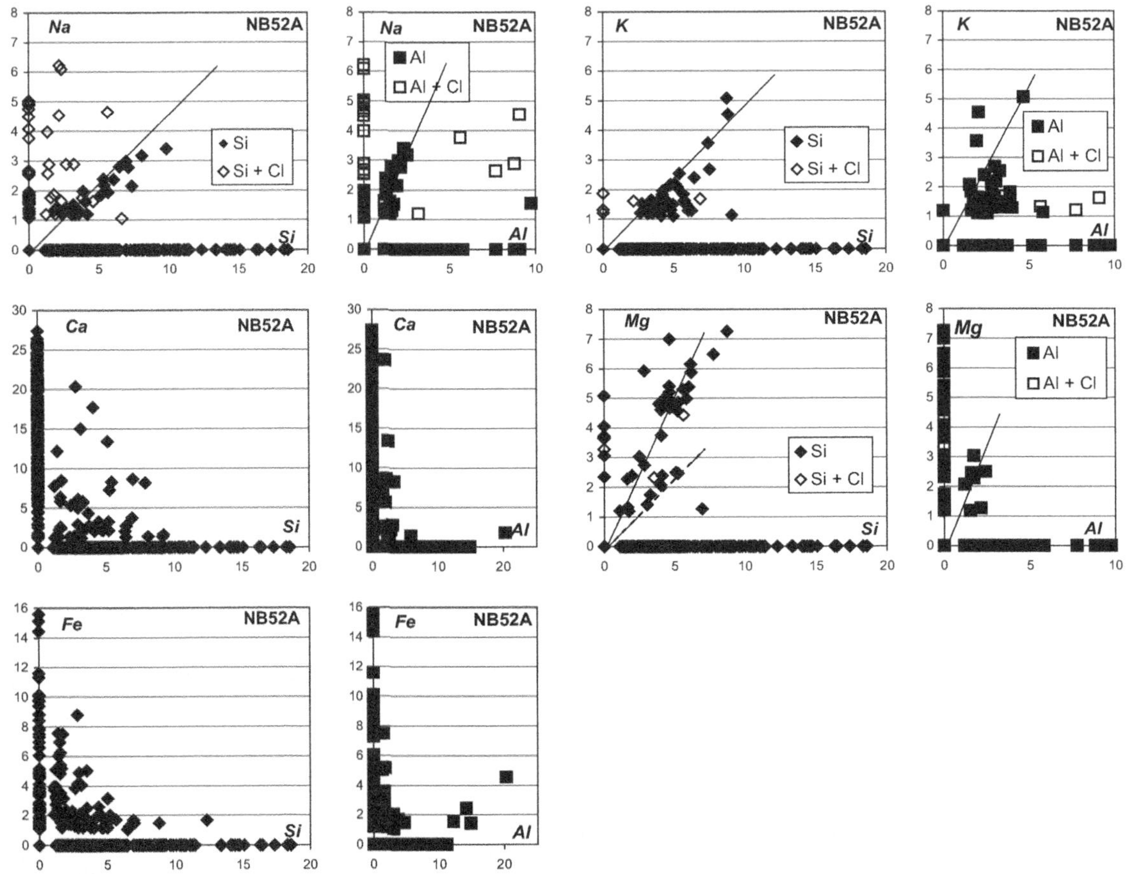

Figure 4.7 Sodium, potassium, calcium, magnesium and iron content as a function of silicon and aluminum content (at%) for particles >2 μm. NB: Si + Cl and Al + Cl: the particle contains also chloride. Lines are drawn with the naked eye

The many observations of Ca on the vertical axis indicate that calcium is predominantly not present as an (alumino-)silicate. The most plausible explanation is that calcium is present as calcite. As nearly all magnesium is present as (alumino)silicates, dolomite ($CaMg(CO_3)_2$) may only be present in very minor amounts. Silicon appears to be present in low contents at high Ca contents, indicating that calcite may contain minor amounts of silicon, or amorphous silica precipitated on calcite during drying of the membrane. Calcium is also present in small amounts at high silicon contents, indicating its presence as a minor constituent in (alumino)silicates or as an artifact calcite precipitate on quartz and/or (alumino)-silicates. Closer inspection of the results shows that calcium is not present as an artifact sulfate or chloride precipitate.

Similarly, the many observations of Fe on the vertical axis indicate that there is no relation between Fe and Si and/or Al. At low contents, iron may be present as a minor constituent in (alumino)silicates. As will be shown later, high iron contents may refer to corrosion and erosion products.

So far, the content of individual cations has been plotted against the content of silicon and aluminum. However in many minerals cations are present in variable proportions. But plotting the sum of the (molar) K, Mg and Fe contents against the silicon and the aluminum content did not yield additional information.

4.3.4 Mineralogical identification of non silicates

In the preceding section it was not possible to relate the greater part of calcium and iron with (alumino) silicates. Considering other possible anions, calcium must be present as calcite (CaCO3). As the presence of magnesium was almost completely explained as (alumino)silicate (Figure Appendix A4.6), the presence of dolomite ($CaMg(CO_3)_2$) may only be minute.

Aluminum

High aluminum contents (>10–20 at%) may indicate the presence of elemental aluminum or aluminum alloy as an erosion/wear product from the well or submersible pump (Figure 4.6 and Appendix Figure A4.2). The presence of aluminum hydroxide (gibbsite) is less likely because the area under the EDS signal would consist of contributions of aluminum and oxygen and would be larger than the area in case of elemental aluminum. As the content of aluminum is calculated as the fraction of the area corresponding with aluminum over the total area (Analysis), the aluminum content in gibbsite is lower than in elemental aluminum. Consequently, a high aluminum content indicates elemental aluminum.

Chromium

Figure 4.8 shows Fe and S as a function of Cr and Ni for SO21. From this figure it is clear that chromium and nickel are present as a minor constituent in iron particles without sulfur, indicating their presence in an iron alloy, which has been eroded from the submersible pump and less likely from the well.

Copper

The highest proportion of particles containing copper occurs in SO21 (Table 4.4). Figure 4.9 shows iron and sulfur as a function of copper and lead. From this figure it may be deduced that copper particles are present as sulfides, whether or not together with lead. Two observations are present at about 1 at% Cu, which contain iron, copper and sulfide. This occurrence as sulfides indicates a provenance as corrosion product from the well, inclusive sample valve, and not to an erosion product.

Iron

Again, both wells of SO show the highest iron contents. Figure 4.10 shows the iron content as a function of the particle diameter. In this figure several iron compositions may be distinguished: iron together with silicon whether or not accompanied by other elements, iron together with sulfur whether or not accompanied by other elements, iron accompanied by some other metal(loid) element (Al, As, Bi, Cr, Cu, Ni, Pb, Sn, Ti, Zn) but without silicon or sulfur, and elemental iron.

This figure shows that the occurrence of iron silicates is limited to low iron contents. Iron sulfides usually show the highest iron content, but some observations fall into the iron silicate field. The presence of iron sulfides not necessarily coincides with the presence of pyrite. Some of them contain lead (see later),

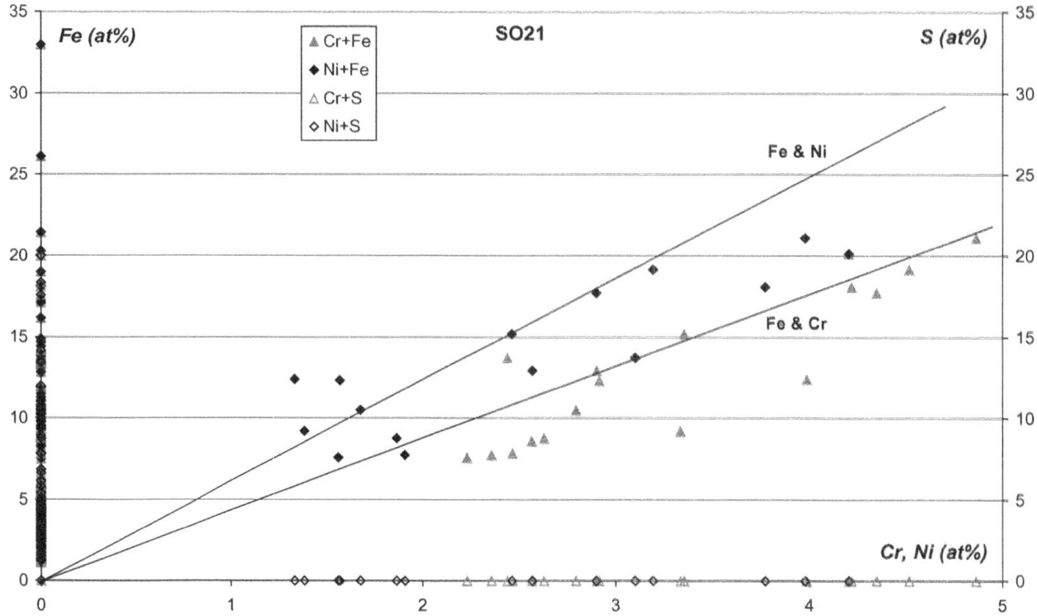

Figure 4.8 Iron and sulfur content as a function of chromium and nickel content for SO21. NB: Lines are drawn with the naked eye

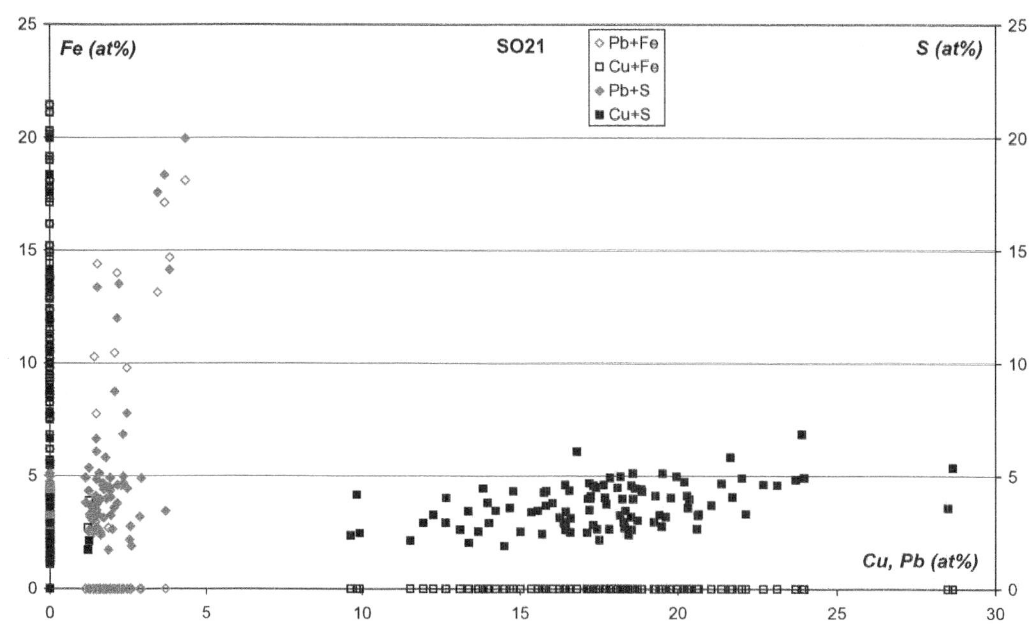

Figure 4.9 Iron and sulfur content as a function of copper and of lead content for SO21

indicating the presence of corrosion products. The fields of iron alloys and elemental iron more or less coincide, indicating that these particles indeed consist of elemental iron or of iron alloy containing <1% of the contributing metals.

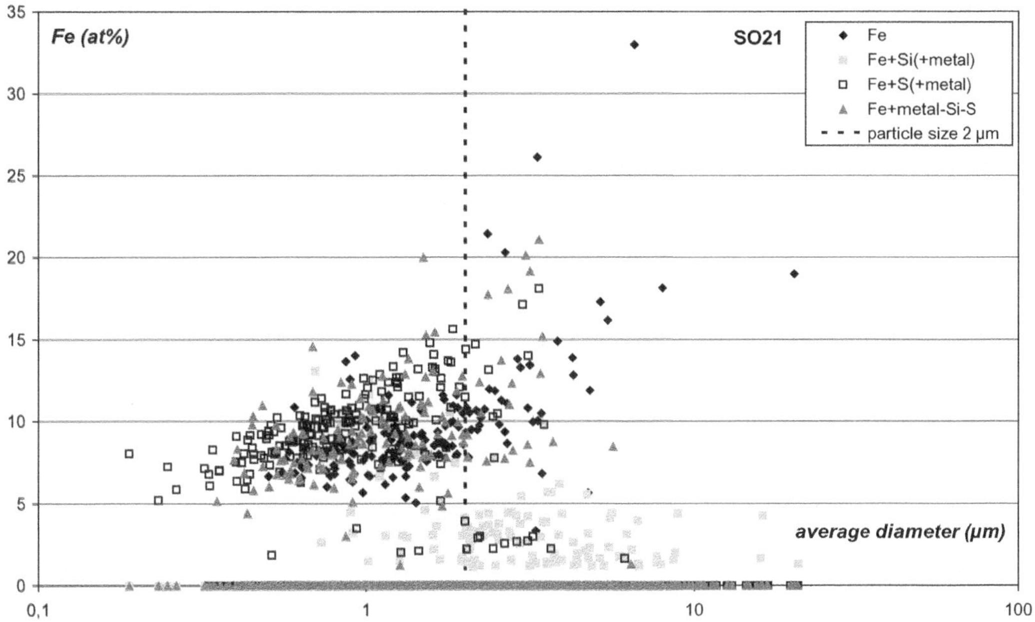

Figure 4.10 Iron content as a function of particle diameter. Four different qualities of iron have been distinguished. Some iron particles contain silicon as well as sulfur

However, the presence of iron particles as hydroxides may not be completely ruled out. Under sulfate reducing conditions iron (hydr)oxides, like hematite and goethite, may be stable (Jakobson and Postma 1999; Appelo and Postma 2006). But these iron hydroxides will not be present as primary minerals, and if present only in very minute amounts. Iron hydroxides are usually present as secondary minerals, as coatings around grains. These secondary minerals usually contain impurities, making them more soluble. Moreover, the solubility is also influenced by the particle diameter: the smaller the particle, the more soluble (Stumm and Morgan 1996; Langmuir 1997). Due to their chemical composition iron hydroxide (Fe_2O_3, $Fe(OH)_3$) particles have a low iron content. If these particles would be present, the "Fe-field" in Figure 4.10 would be divided in two separate "fields", which is obviously not the case. The absence of an "iron (hydr)oxides" field, supported by theoretical considerations, leaves the conclusion that iron hydroxide particles are not present.

Huisman (1998) mentions the presence of siderite ($FeCO_3$), which is stable under methanogenic alkaline conditions (Postma, 1982). For siderite applies the same as for iron (hydr)oxides: the absence of a second "Fe field" implies that siderite particles are also not present.

Lead

The only site containing lead is SO. Figure 4.9 shows the iron and sulfur content as a function of the lead content. From this figure it may be derived that under 6% Fe (and S), lead particles are present as sulfides, generally also together with copper. Above this value lead particles are present together with iron and sulfide. These compositions indicate a provenance as corrosion product from the well, inclusive sample valve, and not to an erosion product.

Nickel

Nickel only occurs together with iron (Figure 4.8), and has already been discussed in relation with chromium.

Natural iron sulfides usually contain admixtures of cobalt, nickel, zinc and arsenic (van Beek et al. 1989; Kjoeller et al. 2004), where the low contents of cobalt and nickel are often comparable. As cobalt has not been encountered, and nickel is only present together with iron and chromium and not with sulfides, the presence of natural sulfides could not be confirmed.

Zinc

Particles containing zinc are most often found in VE14. Figure 4.11 shows for VE14 iron and sulfur as a function of some trace elements. Zinc occurs only together with sulfur. Apart from bismuth, zinc sulfide contains no other trace elements; this "purity" indicates an origin as a corrosion product, and not to a natural provenance.

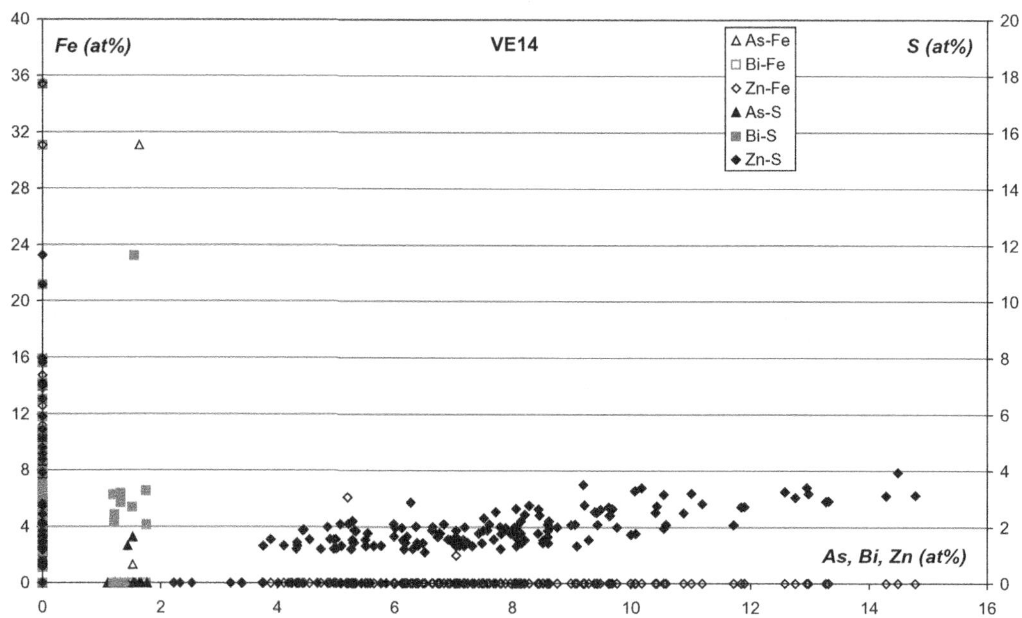

Figure 4.11 As, Bi and Zn as a function of Fe and S for VE14

Other elements

For the presence of As, Ba, Bi, Mn, Sn, Sr and Ti see Appendix Paragraph A-1.

4.3.5 Mineral frequency distribution

Artifact particles

During drying of the membranes all material dissolved in the adhered moisture precipitated, as exemplified by the presence of halite. This amount will be proportional to the volume evaporated and to the concentration (Table 4.3).

On all sites, except SO, abstracted groundwater is saturated with respect to calcite (Table 4.3). Thus immediately upon evaporation calcite will start to precipitate as a film on the filter membrane. This phenomenon is comparable to the precipitation of salts in a playa lake (Drever 1988). Soon, the solution will also become saturated with respect to amorphous silica, and this mineral will also precipitate as a film on the membrane. Upon further evaporation, due to their greater solubility, sulfates, carbonates (except calcium) and chlorides will reach high concentrations. These high concentrations may precipitate as separate salt particles or under influence of capillary forces move to filtered particles, and subsequently precipitate around these particles, together with calcite and silica.

Not all artifact precipitates may have been recognized. Moistening polycarbonate membrane filters in the laboratory in triplicate yielded an average thickness of the water film of 0.11 mm. Assuming a hardness of 2 mmol/l (Table 4.3), a molar weight of 100 g, and a specific density of 2.5 g/cm^3, results in an average thickness of the calcite precipitate of 0.009 μm. This film is too thin to be recognized by EDS (Reimer 1998).

Mineral frequency distribution

In the preceding section, based on the elemental composition, each particle was assigned to a mineral. Here, with the help of a decision tree (Figure 4.2), all particles >2 μm are classified into broad mineral groups (Table 4.5). These results were corrected for the presence of artifact (chloride) precipitates. Only on rare occasions there was more than one mineral, i.e. stacking. This number ranged from 0 in RH06, TW33 and VE22 to 5 in SO21. The difference in total number between Table 4.5 and Table 4.4 equals the number of particles consisting of organic matter and salt (halite) precipitates.

Table 4.5 Mineralogical distribution of particles >2 μm (%)

	NB50B	NB52A	RH04	RH06	SO21	SO22	TW22	TW33	VE14	VE22
total number	741	673	182	554	658	660	1065	689	303	205
quartz	20.5	19.7	26.2	45.1	42.1	40.2	18.9	28.0	13.5	6.3
cation silicates	20.9	31.8	37.7	25.6	5.9	5.4	31.1	40.5	40.4	62.0
alumino silicate	24.8	19.7	3.3	1.2	13.0	22.4	17.6	9.9	3.4	1.3
cation alumino silicates	33.8	28.8	32.8	28.0	39.0	32.0	32.4	21.6	42.7	30.4
total silicates	100	100	100	100	100	100	100	100	100	100
silicates	40.8	50.5	67.0	14.8	59.6	79.1	6.9	57.0	29.4	38.5
calcite	44.0	37.4	21.4	82.3	11.1	1.2	91.8	37.3	44.2	53.7
iron sulfide	0.8	1.9	0	0.4	1.2	0.6	0.1	0.3	0	0
other sulfides	0.1	0	0.5	0	16.6	3.8	0.4	0	20.5	0.5
metals	14.3	10.1	11.0	2.5	11.6	15.3	0.8	5.4	5.9	7.3
total minerals	100	100	100	100	100	100	100	100	100	100

The results of Table 4.5 show large differences in mineral composition: in SO predominantly silicates and in particular quartz and clay minerals, and in RH06 en TW22 predominantly calcite. The other wells show various proportions of silicates (inclusive quartz) and calcite. Remarkable are the low proportions of kaolinite in RH and VE. Possible explanations for large differences in mineral distribution between wells from the same well field, may be the abstraction from different aquifers (Table 4.1: TW). Table 4.5 also shows that iron sulfide particles are hardly present, and that the proportion of particles due to corrosion and erosion usually varies between 1 to 10%, but may run up to more than 25% as in SO21 and VE14.

Striking are the differences between the major cations (Appendix Figure A4.3 to A4.7 inclusive): Na and K almost completely present as alumino silicates, like feldspars, Mg more as silicates than alumino silicates, Ca more as silicates than alumino silicates, and Fe the opposite, more as alumino silicates than silicates.

Remarkable is the proportion of calcite particles. This contribution varies from 1% in SO22 to 92% in TW22, while the calcite content of the aquifer matrix varies from <0.1 wt% (calcite absent) in NB and SO to over 3 wt % in VE. As the specific weights of quartz (≈2550 kg/m^3) and calcite (≈2700 kg/m^3) are comparable, calcite particles are overrepresented with respect to the aquifer matrix composition. Indeed, minerals are not equally divided over the various grain sizes (Zonneveld 1947; Pettijohn et al. 1987), decreasing grain sizes show decreasing quartz and increasing clay mineral contributions (Breeuwsma 1990).

The presence of precipitates like chlorides, calcium carbonate and silica together with corrosion particles may be used to obtain information about the accuracy of the mineral distribution in Table 4.5. If chloride occurs together with corrosion particles, the accompanying cations may be considered salt precipitates from drying, and if calcium or silicon occur together with corrosion particles, these occurrences may be considered as precipitates or as stacked calcium or silicon minerals. Figure 4.12 shows the chloride, calcium and silicon content as a function of the sulfide content for SO21 and VE14, containing the highest proportion of corrosion particles with respectively hardly any and many chloride precipitates (Table 4.4). This figure shows that in SO21 on 126 corrosion particles respectively 8 particles contain calcium and 17 particles silicon, and in VE14 on 82 corrosion particles respectively 9 particles contain calcium and 10 particles silicon. Consequently, roughly 10% of the filtered particles seem involved in precipitation and/or stacking.

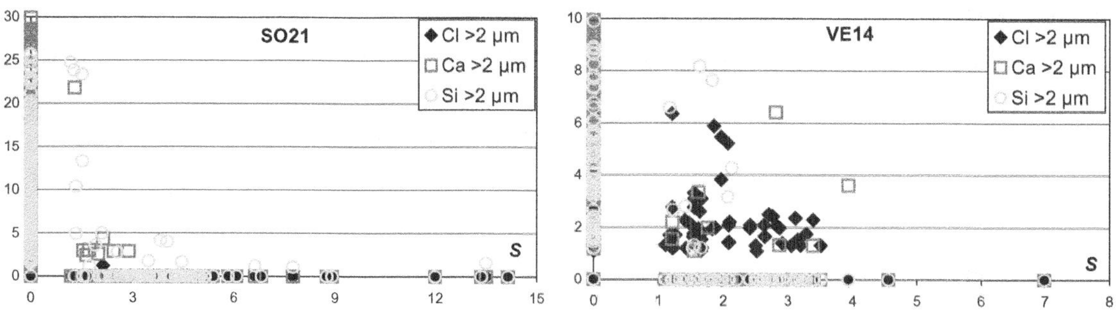

Figure 4.12 Chloride, calcium and silicon content as a function of S content (at%) for particles >2 µm in SO21 and VE14

Assuming absence of preference in substratum for precipitation of calcium carbonate or amorphous silica, the accuracy of Table 4.5 is estimated at circa 10%. Precipitation and/or stacking of chlorides or silicates on calcite particles results in an underestimation of calcite, and in an overestimation of cation silicates. Precipitation and/or stacking of chlorides and calcite on quartz particles results in an overestimation of cation silicates at the expense of quartz, and on alumino silicate particles in an overestimation of cation alumino silicates at the expense of alumino silicates. Consequently, the proportion of calcite particles may be even somewhat higher than presented in Table 4.5. Chloride and calcium carbonate precipitates only affect the distribution within the silicates. The rather high aluminum cut off value has the same effect.

4.4 DISCUSSION

In our research, in anoxic (sulfate reducing) groundwater no manganese oxide or iron (hydr)oxide particles were observed, while in literature these particles are frequently reported. This is in line with the data of Ryan and Gschwend (1990) who analyzed the mineral composition of colloids present in acidic (pH 3.9–6.3) groundwater obtained from observation wells during minimal drawdown. Each site comprised two wells; one well screened in oxic groundwater (containing oxygen and nitrate) and the other well in anoxic groundwater (containing iron and sulfate) of the same geological formation. The colloids under oxic conditions consisted of redox insensitive minerals only like kaolinite and muscovite, whereas under anoxic (iron reducing) conditions they consisted of redox sensitive (possibly $Fe(OH)_3$ and goethite) and insensitive minerals (quartz, possibly gibbsite, kaolinite, chamosite, and muscovite). The low pH of the overlying groundwater reflected the absence of calcite in these deposits, and consequently the absence of calcite particles.

This is in line with theory; depending on the availability of reducing material (organic matter) in the aquifer, during flow groundwater will become increasingly reduced. Under continuously more reducing conditions manganese oxide and iron (hydr)oxide particles will dissolve, and iron sulfide particles will be formed. Consequently the mineral composition of particles reflects the groundwater chemistry, as controlling their stability.

Particles may be supplied to groundwater by dissolution of minerals or by in situ precipitation of supersaturated mineral phases (Ryan and Elimelech 1996; Kretzschmar et al. 1999). By dissolution, in particular of (hydr)oxide and carbonate coatings on grains or of cement between grains, enclosed minerals may be released (Reed 1977; McCarthy and Degueldre 1993; Hofmann et al. 2003), for instance clay minerals (Ryan and Gschwend 1990) and silicates (Gschwend et al. 1990).

Existing particles may also become mobilized by chemical perturbation (Ryan and Elimelech 1996; Kretzschmar et al. 1999). Such a perturbation process involves the presence of intercalated clay layers in aquifers. In these clay layers groundwater chemistry is quite different (more reducing) from groundwater in the surrounding aquifer (Brown et al. 1999, 2000). Groundwater leaving these clay layers and entering the sandy matrix of the aquifer experiences a chemical shock, mobilizing particles at hand, like clay minerals and organic matter (Blume et al. 2002).

Consequently the mineral composition of particles reflects not only the groundwater chemistry but also the aquifer mineralogy, as a source of particles.

Most surprising of the mineral particle composition compared to the sediment grain composition is the overrepresentation of calcite. This might be explained by differences in particle size or particle shape. Figure 4.13 shows the particle diameter frequency distribution of calcite and quartz of the two most contrasting sites: SO21 with the highest frequency of quartz, and TW22 with the highest frequency of calcite particles. In both cases calcite shows the greater average diameter. Consequently, the predominance of calcite over quartz particles cannot be explained by a difference in diameter. However, there are sites which show the opposite situation: quartz particles being larger than calcite particles.

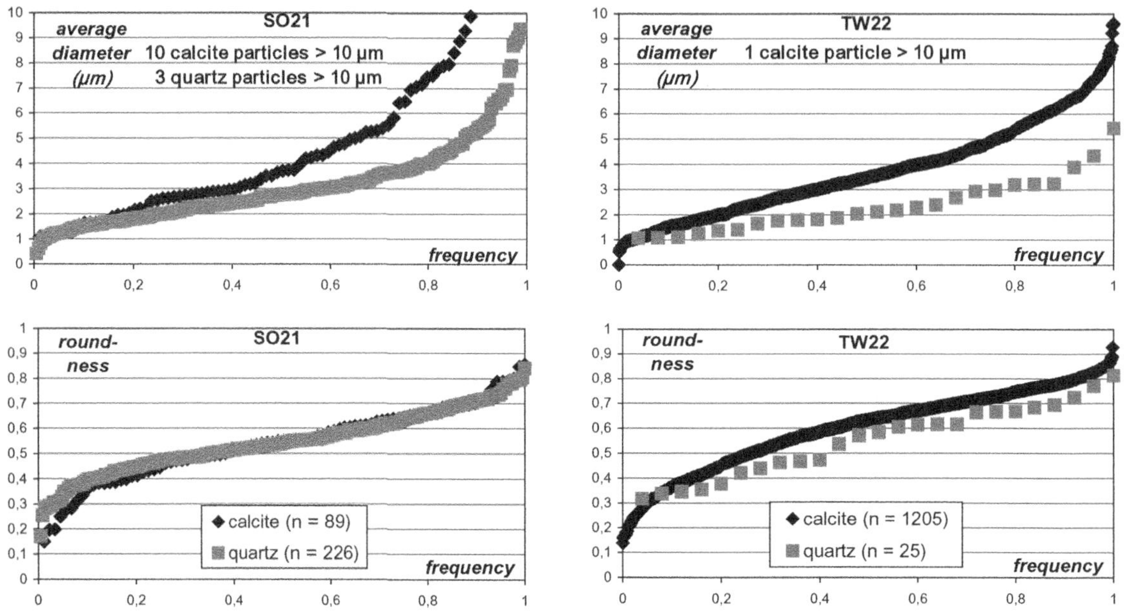

Figure 4.13 Frequency distribution of particle diameter and roundness of calcite and quartz for SO21 and TW22

The shape of particles may be expressed in various ways; here the shape is expressed as the roundness: $4A/(\pi d_{max}^2)$ with A measured area and d_{max} measured maximum diameter. Figure 4.13 shows that there are hardly any differences in shape between calcite and quartz, which might explain the predominance of calcite over quartz particles.

Except for SO, at all other sites the confining layer above the upper aquifer contains calcite (van Duijvenbooden and Breeuwsma 1987). As soil calcite is generally readily soluble, ground water will quickly become calcite saturated. Due to later redox processes producing alkalinity, groundwater will become supersaturated with respect to calcite. In soils calcite precipitates as concretions and as coatings on soil matrix grains (Schachtschabel et al. 1989) but probably also as particles. Continuous calcite precipitation will occur in particular in NB and RH because these sites show active sulfate reduction and groundwater is calcite saturated. Nevertheless the small negative value for the Saturation Index (Table 4.3), due to uncertainties in the calculation $SI_{calcite} = 0 \pm 0{,}1$ is considered calcite saturated (Appelo and Postma 1996). This explanation is less likely for SO, TW and VE, as at these sites sulfate reduction is largely finished. Generation of particles due to super saturation has been described for phosphate (Siemens et al. 2004; Stollenwerk 1996; Gschwend and Reynolds 1987).

Methanogenesis produces besides methane also carbon dioxide, which might (locally) lead to calcite under saturation. If so, smaller calcite particles will dissolve preferably over larger ones (Stumm and Morgan 1996). Consequently, methanogenesis should lead to a decrease of calcite particles, but as the greater part of the particles consists of calcite, this process is apparently of negligible importance.

Sand deposits usually contain a trace of fine material (1à2% <2 µm and 1à2% between 2 and 16 µm; Pettijohn et al. 1987; van der Sluijs and Locher 1990). In fluvial sediments these particles cannot be deposited together with these sands, but must have been infiltrated later. These particles consist of standard sedimentary material like quartz, clay minerals, feldspars, micas, etc., but not of calcite as these particles will dissolve.

Transport of particles through fine sands is very limited (Harvey et al. 1989; Bradford et al. 2002; Close et al. 2006). Considering the low groundwater flow velocities in the Netherlands (order of magnitude: Darcy flow velocity circa 35 m/year) and the long distances covered, which are equivalent with the residence times in Table 4.2, one would expect abstracted groundwater to be devoid of (calcite) particles. But, surprisingly, the greater part of the particles consists of calcite. This presence indicates a local production, which is corroborated by large differences in particle mineral composition between both wells on one site (RH). As all sites show the same overrepresentation of calcite over quartz particles, at all sites the same process must be effective.

The same reasoning more or less applies to the presence of quartz particles. On all sites abstracted groundwater is supersaturated with respect to quartz (Table 4.3) (solubility at 25°C 0.1 mmol/l; Stumm and Morgan 1996), but no quartz particles will develop as this process is very sluggish (Appelo and Postma 1996). As the groundwaters are not saturated with respect to amorphous silica (solubility at 25°C 2 mmol/l; Stumm and Morgan 1996), no such deposits (particles) will develop.

As the overrepresentation of calcite particles in, at least, TW and VE cannot be explained by chemical and sedimentary processes, there must still be another process. Such a process might be a mechanical driving force: crushing or abrasion of calcite and quartz grains and/or coatings present in the soil matrix: quartz being much harder than calcite, is much more resistant to crushing or abrasion. Crushing and abrasion may occur by the weight of the overlying deposits, and will be promoted by vibrations. Vibrations may occur as minor, unnoticed, earth quakes (Brodsky et al. 2003) and, close to the well, by switching on and off the (submersible) pump.

Clay mineral and organic matter particles may also be generated in other ways. Jongmans (1999; 2004) remarked the presence of clay balls in aquifer deposits in NB and RH. These balls will easily crumble upon increasing pressure, while generating many particles.

4.5 CONCLUSIONS

4.5.1 Particle mineralogy

Automated SEM-EDS is a very useful method to analyze large numbers of particles, yielding representative elemental and morphological information.

By comparing and combining the relative contents of the various elements of the SEM-EDS analysis it is possible to identify specific minerals like quartz, sodium feldspar, potassium feldspar, muscovite, biotite, clay minerals, calcite and iron sulfide. Mineralogical methods have not yet been applied to confirm this chemical identification.

In our research no manganese oxide or iron (hydr)oxide particles were found. This is in line with the sulfate reducing nature of the abstracted groundwater, under which conditions these secondary (hydr) oxides are not stable. Consequently, particle mineral composition is governed by aquifer mineralogy, as source, and by groundwater chemistry, as controlling the stability of particles.

With the help of a procedure (decision tree), the particle mineral frequency distribution was estimated from the elemental composition. There appear to be large differences between the well fields, and sometimes also between both wells of one well field. The proportion of calcite particles varies between 1 and 92% of the total number, and of silicates between 7 and 79%. The silicates consist mainly of quartz, cation silicates and cation alumino silicates; alumino silicates (kaolinite) usually contributing the least. The contribution of iron sulfide particles (<2%) is negligible.

The mineral particle frequency distribution is not proportional to the mineral grain frequency distribution of the aquifer soil matrix. The aquifer matrix contains up to circa 3 wt% of calcium carbonate, while the particle number frequency goes up to 92%. With respect to the soil matrix mineralogy, calcite particles are thus overrepresented. This overrepresentation cannot be explained by differences in size or shape, facilitating transport through the matrix.

Surprisingly, between 2 to 10%, but even up to 25%, of the particles consists of copper, lead and zinc sulfides and of metals like elemental iron, whether or not mixed with chromium and nickel, and aluminum. These particles are considered corrosion products from the well and sample valve and erosion products from the submersible pump.

4.5.2 Particle generation

Equilibrium between groundwater chemistry and particle mineral composition, overrepresentation of calcite over quartz particles, and large differences in particle mineral distribution of nearby wells, all indicate local particle generation.

The generation of calcite particles cannot be explained by dissolution, as groundwater is generally already saturated with respect to calcite at shallow depth. Upon redox processes producing alkalinity, groundwater may become super saturated with respect to calcite, and calcite may precipitate as concretions or as skins around grains, but also as particles. As soon as no alkalinity is produced anymore, no calcite particles will be generated along this way. Groundwater is also supersaturated with respect to quartz, but this precipitation process is too sluggish to be of relevance. Sand deposits also contain traces of material $<2\,\mu m$ consisting of standard minerals, but not of calcite. This leaves mechanical generation, i.e. crushing and abrasion of (skins of) grains by the weight of the overlying soil column. Crushing may be favored by small vibrations, like very small earthquakes or by starting (submersible) pumps or by settling of the bore hole and surrounding aquifer after well construction. This reasoning also explains the overrepresentation of calcite particles over quartz and silicate particles: quartz and silicates are much harder than calcite and consequently will be more resistant to fragmentation, while calcite particles are also generated by chemical precipitation.

Clay mineral and/or organic material particles may be mobilized mechanically by "fragmentation" of clay balls or chemically. The chemical composition of groundwater within intercalated clay layers or clay balls is quite different (more reduced) from groundwater in the surrounding aquifer. Upon leaving this more reduced environment, groundwater experiences a chemical "shock", which may mobilize particles.

The results of this study provide information on the mineralogy of particles in abstracted groundwater and a novel mechanism for their generation.

APPENDIX

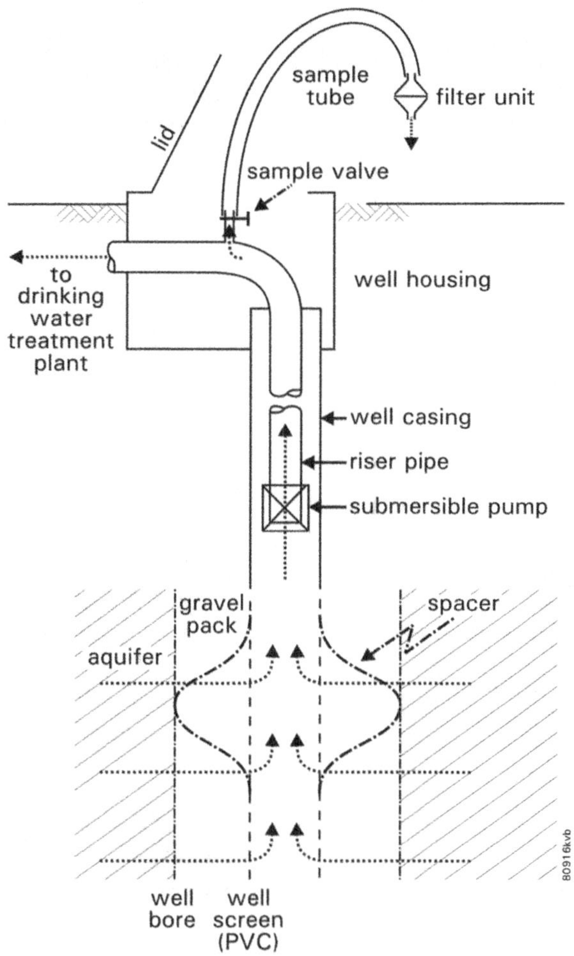

Figure A4.1 Sampling of abstracted groundwater for mineral particle analysis, with route of sampled water indicated

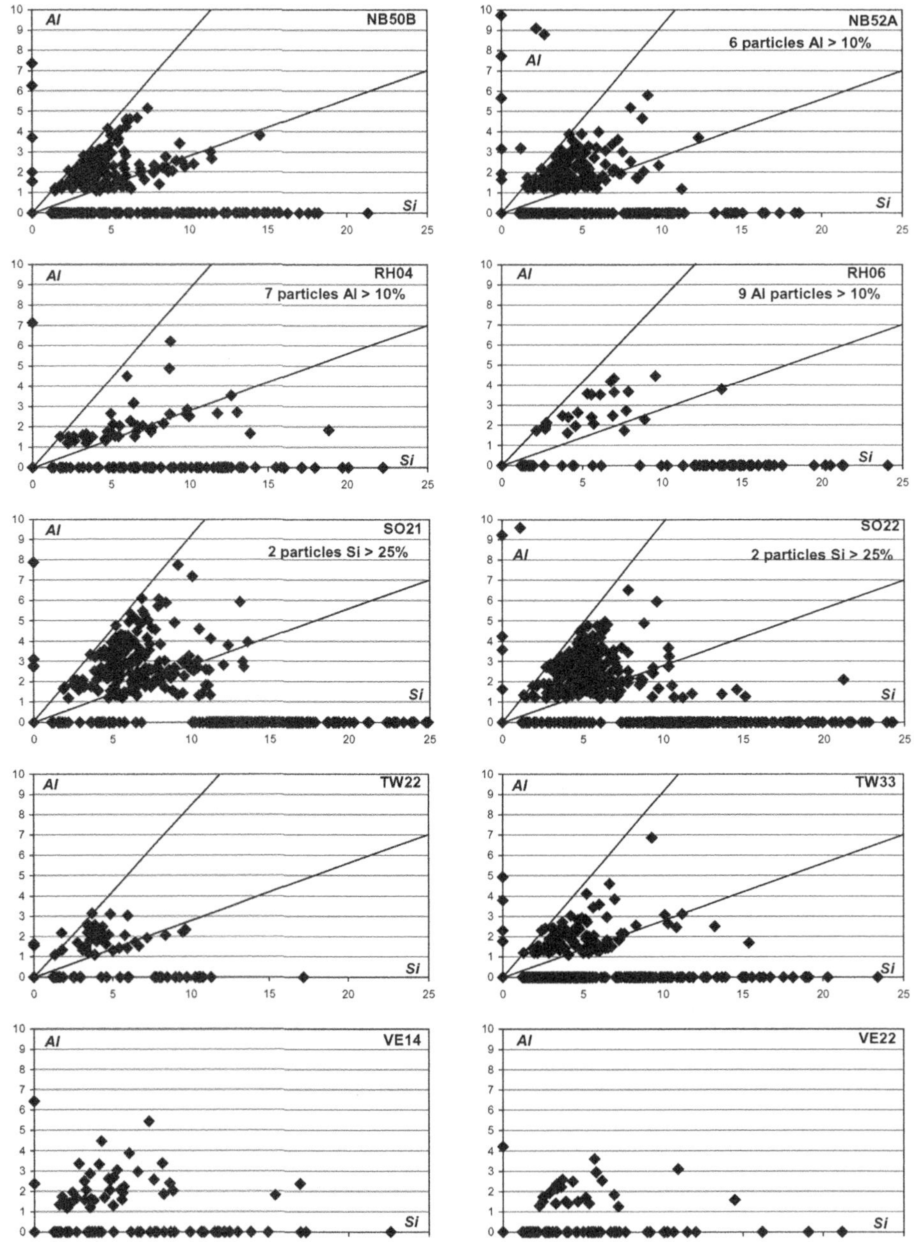

Figure A4.2 Aluminum content as a function of silicon content, in at%, for particles >2 µm. Observations with >10% Al or >25% Si are omitted, but indicated. Lines are drawn for mutual comparison

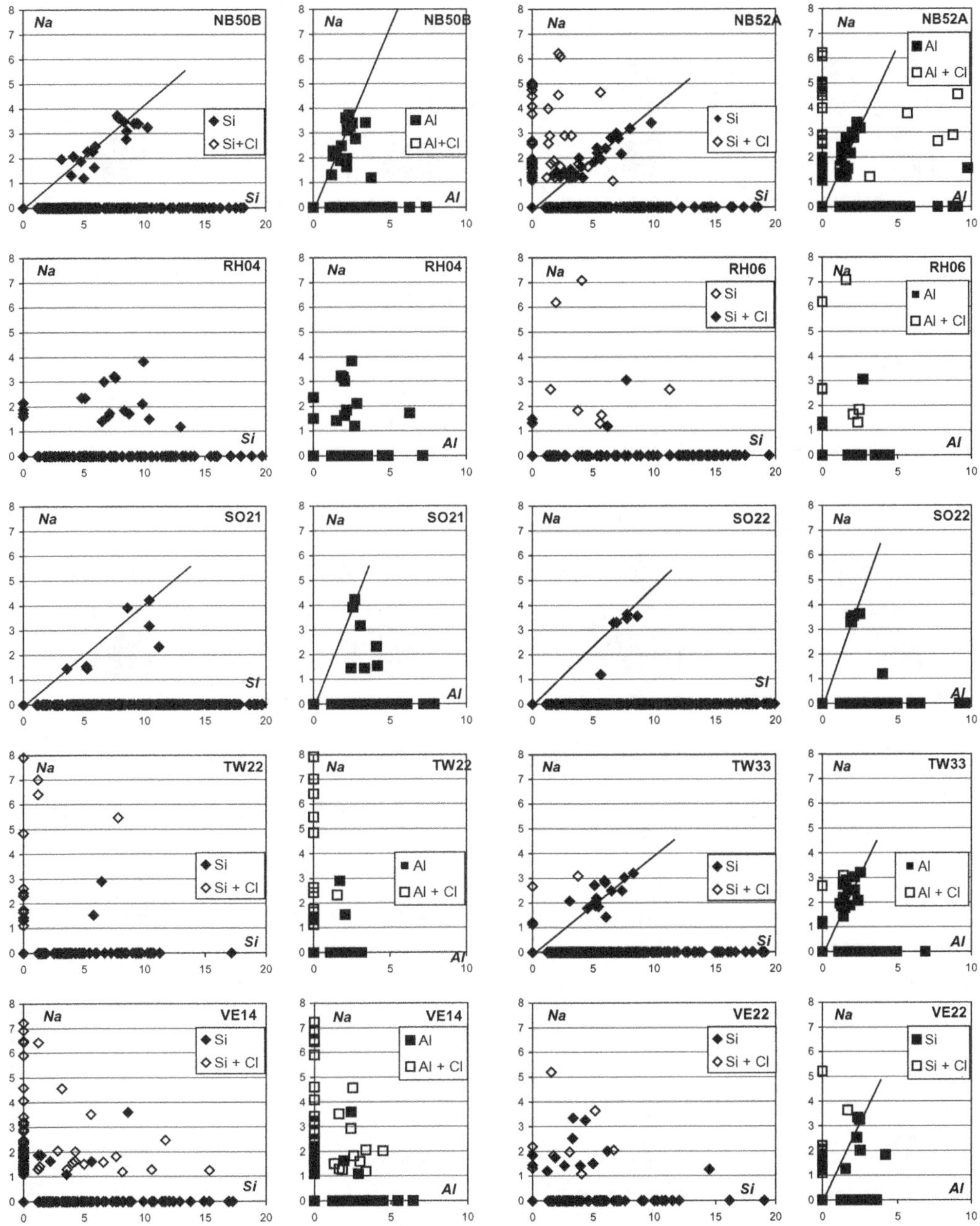

Figure A4.3 Sodium content as a function of silicon content and aluminum content, in at%, for particles >2 μm. NB: Si+Cl: sample contains besides the indicated Si content also Cl, ditto Al+Cl. Lines are drawn with the naked eye

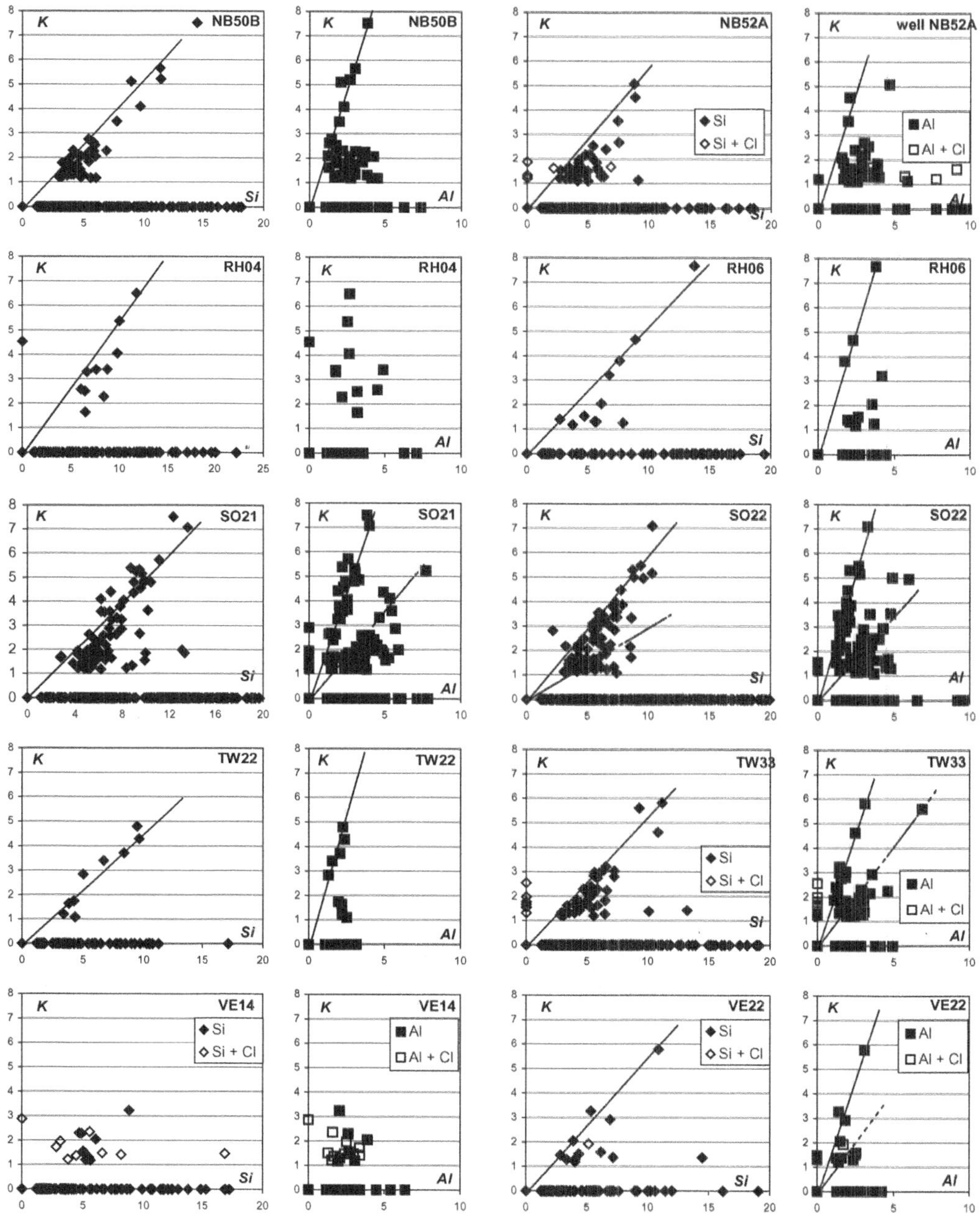

Figure A4.4 Potassium content as a function of silicon content and aluminum content, in at%, for particles >2 μm. NB: Si+Cl: sample contains besides the indicated Si content also Cl, ditto Al+Cl. Lines are drawn with the naked eye

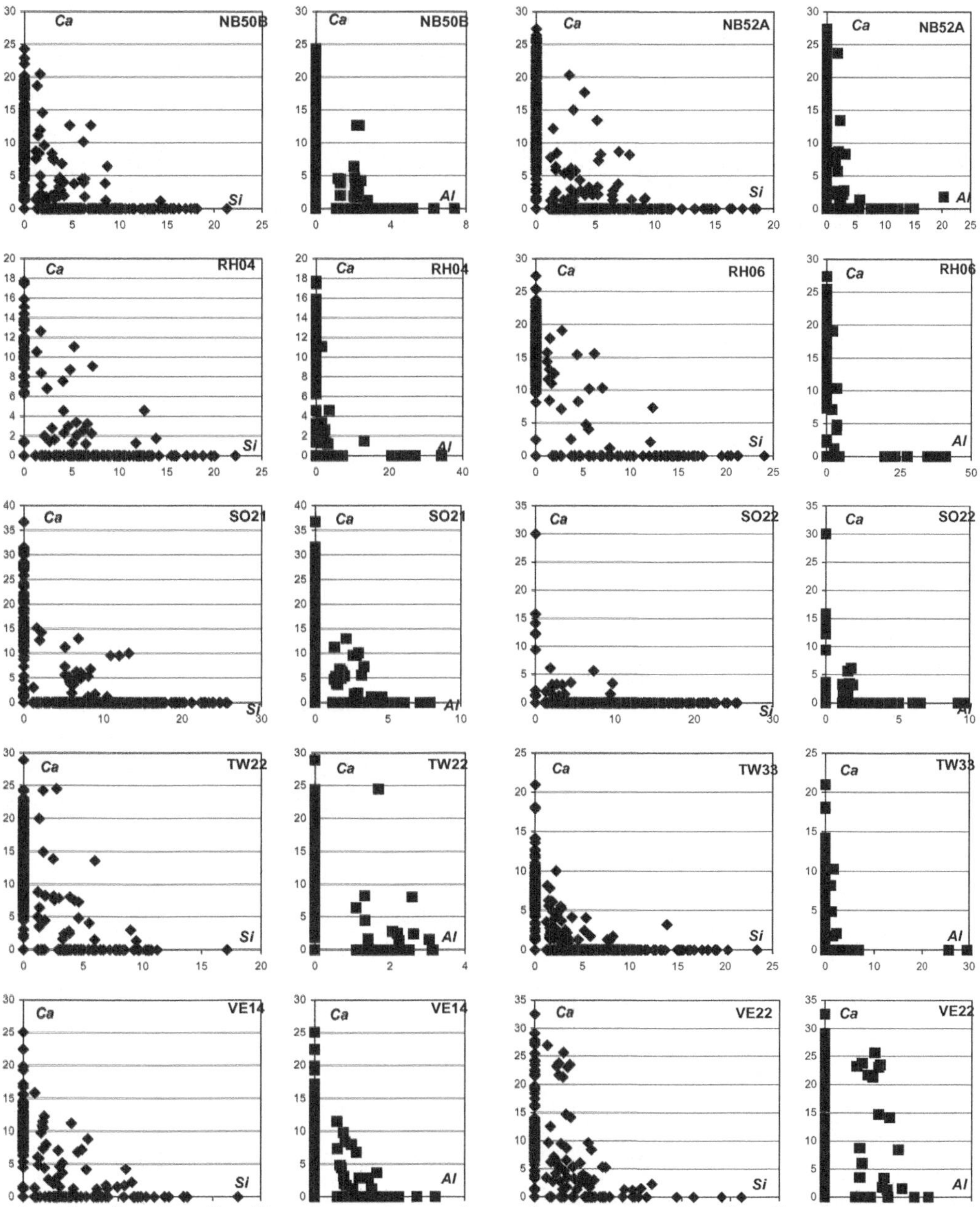

Figure A4.5 Calcium content as a function of silicon content and aluminum content, in at%, for particles >2 µm

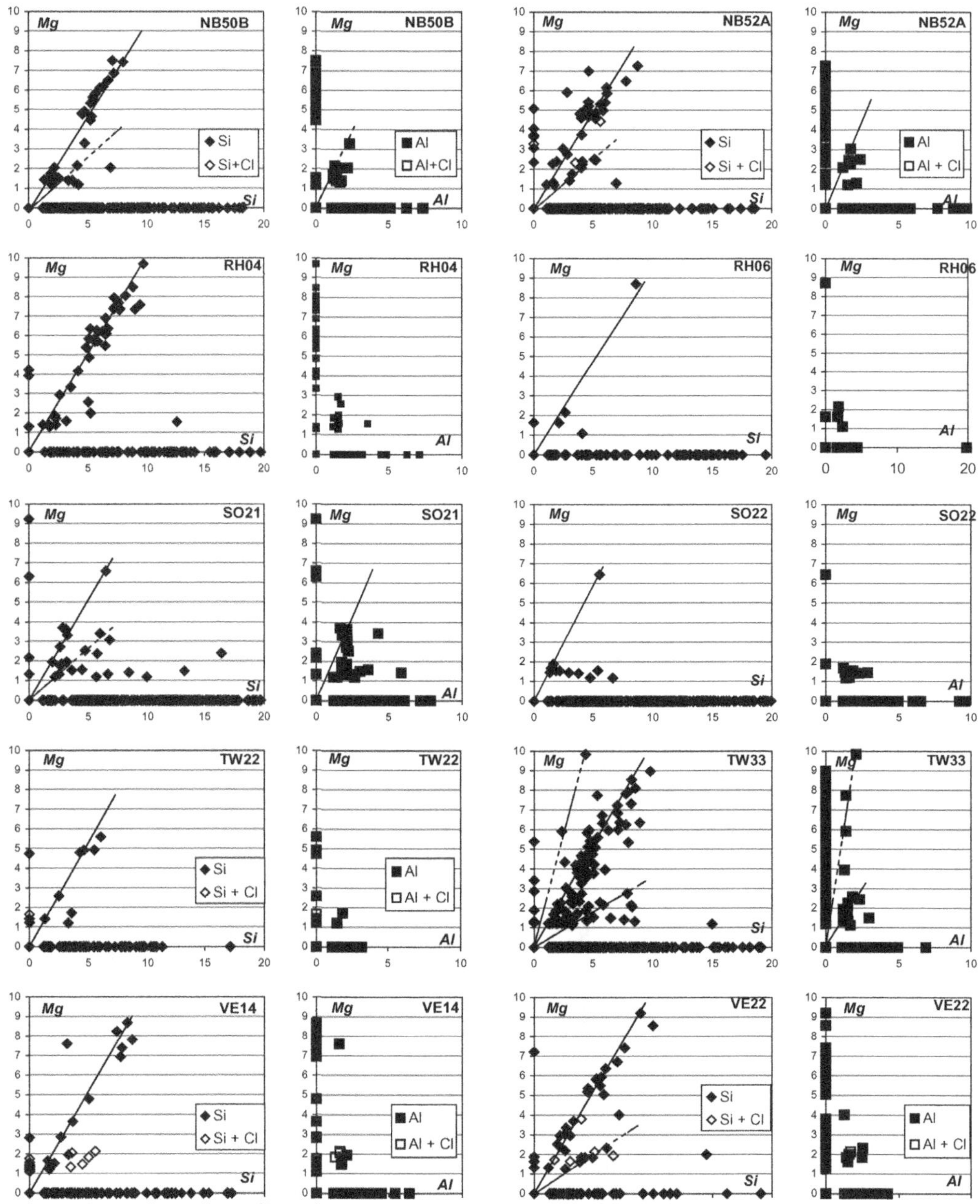

Figure A4.6 Magnesium content as a function of silicon content and aluminum content, at%, for particles >2 μm. Lines are drawn with the naked eye

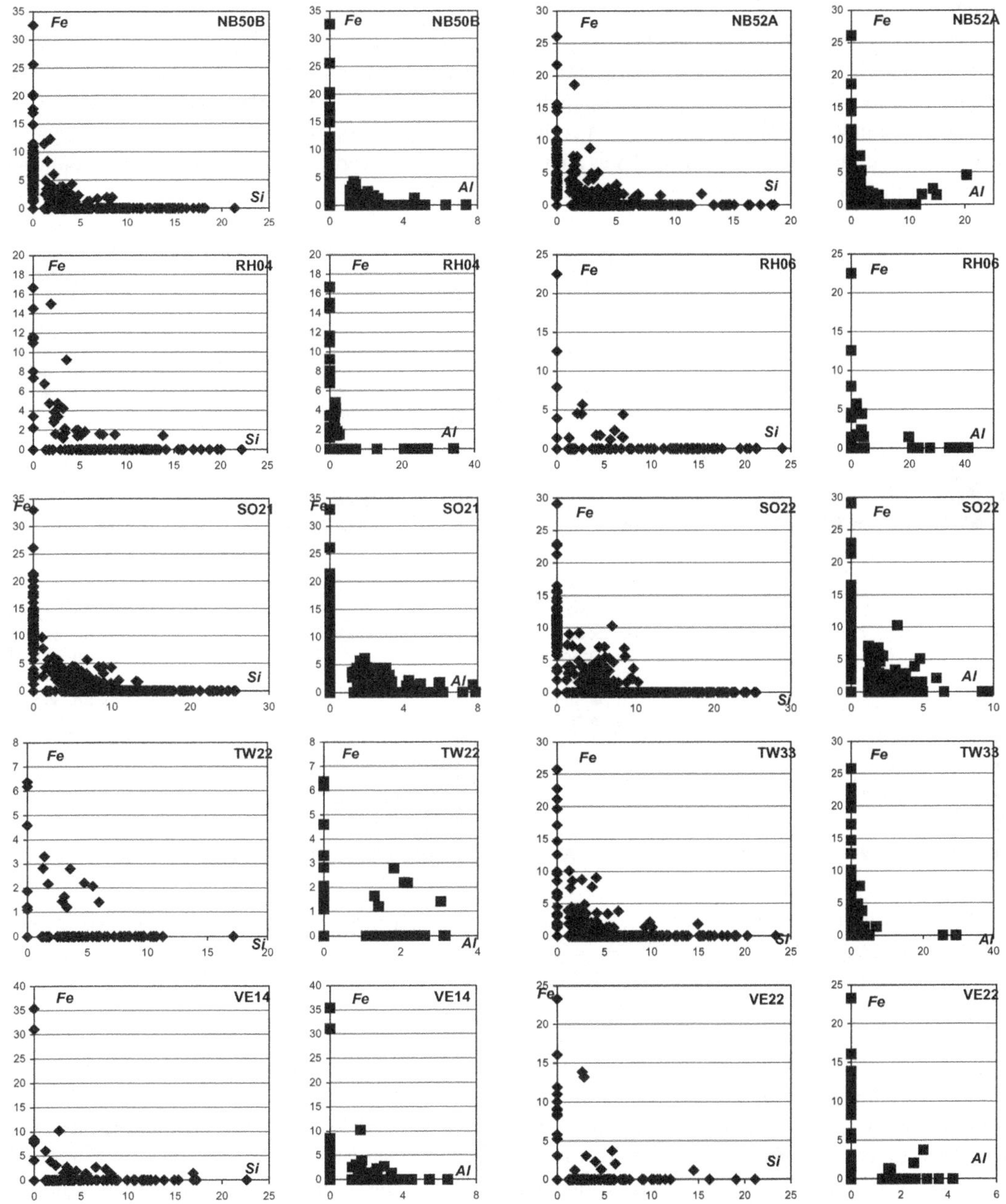

Figure A4.7 Iron content as a function of silicon content and aluminum content, in at%, for particles >2 µm

Paragraph A-1

The presence of trace metals may be related to (Huisman 1998) the presence of phyllosilicates, glauconite (As, Cr, Ni, Pb and Zn), pyrite (As) or iron hydroxides (As, Ba, Ni and Zn), to interaction with organic rich subsurface material (Co, Ni and Zn) or as a result of anthropogenic pollution of topsoils (Cu, Pb and Zn). Figure P4.1 shows the content of the indicated trace metals as a function of Al as proxy for phyllosilicates, K for glauconite, S for pyrite, and Fe for iron hydroxides for NB52A as an average site. However as already observed, glauconite seems to be absent at the sites of this study (Figure A4.4).

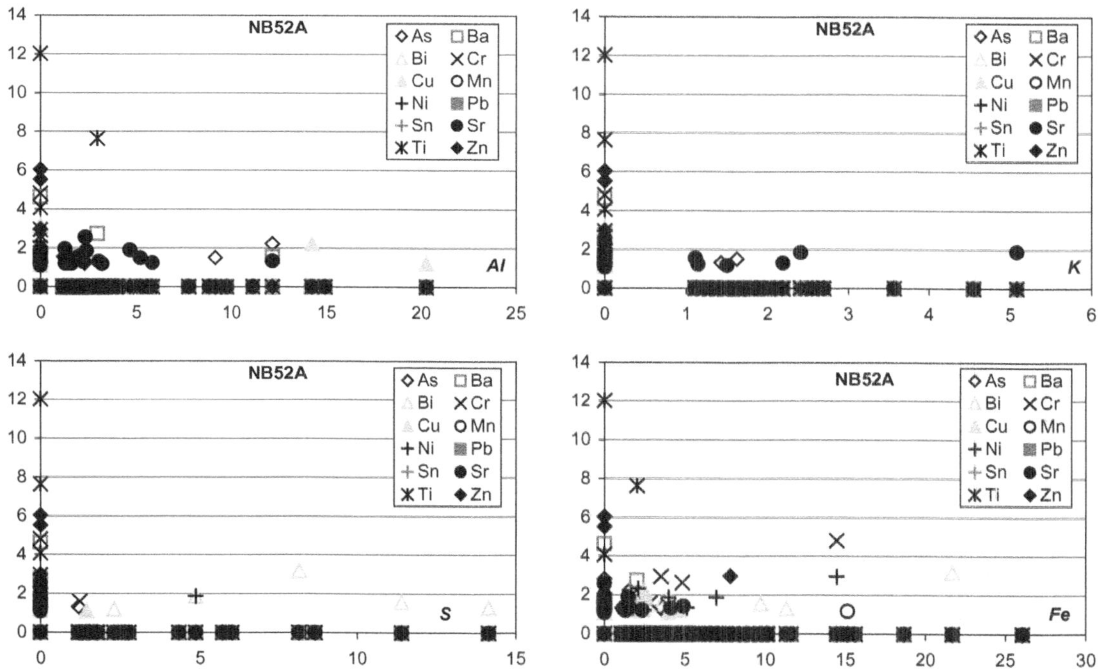

Figure P4.1 Content of minor elements As, Ba, Bi, Cr, Cu, Mn, Ni, Pb, Sn, Sr, Ti and Zn as a function of Al, K, S and Fe content in particles >2 μm in NB52A, all in at%

Arsenic

In Figure P4.1 arsenic is present together with aluminum, potassium, sulfur and/or iron as a minor constituent in potassium-iron-alumino silicates and in iron sulfides.

Both wells of SO contain the highest occurrence of arsenic. Figure P4.2 shows for SO21 the content of various elements as a function of the arsenic content. If there are more observations related to one As value (several observations on one vertical), they belong to the same sample. This figure shows that arsenic is present in various combinations together with calcium and silicon, calcium and iron, copper and sulfur, or together with copper, lead and sulfur, referring to natural as well as artifact particles.

Barium

In Figure P4.1 barium is present together with aluminum and/or iron as a minor constituent in iron-aluminum-silicates. Inspection showed that barium is not present together with sulfate.

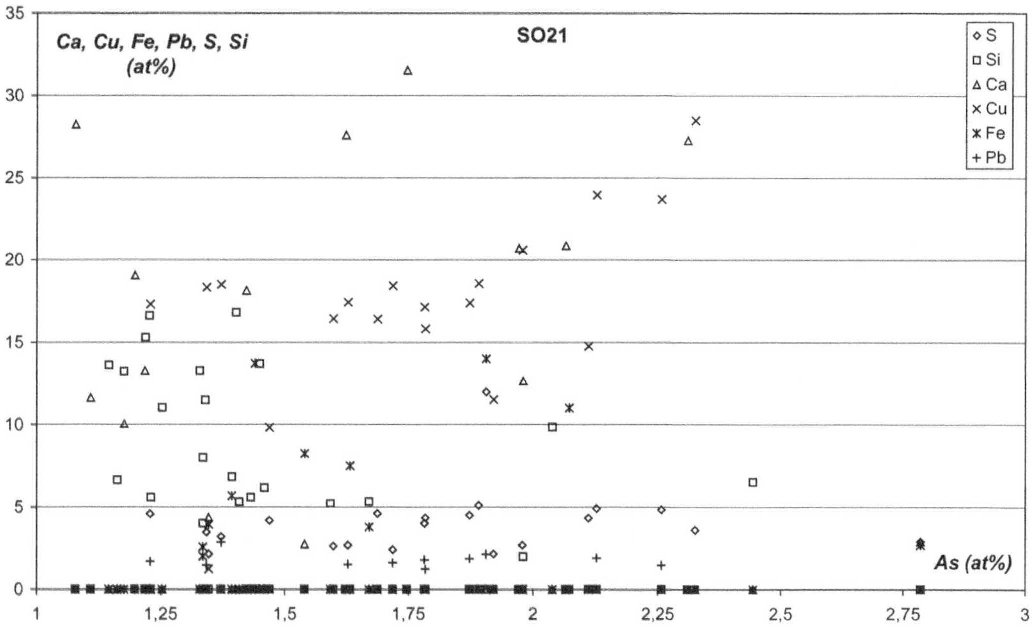

Figure P4.2 Content of various elements as a function of arsenic content for SO21

Bismuth

In Figure P4.1 bismuth is present together with iron and most often also with sulfur. The presence of bismuth together with iron indicates the presence in an iron alloy as an erosion or wear product, and with iron and sulfur as a corrosion product from the well or submersible pump.

Manganese

Manganese is hardly present, and if so only in minor amounts together with iron, as in NB52A.

Strontium

Strontium seems to be present in several shapes: as a minor constituent in potassium-iron-alumino-silicate. However, no relation with calcium was observed.

Tin

Tin is hardly present, only once together with iron in SO21.

Titanium

Titanium occurs as a minor constituent in alumino silicates, whether or not together with potassium and iron as biotite. However, titanium may also occur together with iron as an alloy as an erosion or wear product, or on its own or together with quartz as rutile (TiO_2).

SO shows the highest occurrence of titanium. Figure P4.3 shows for SO21 iron, silicon, barium and aluminum as a function of titanium. This figure shows that titanium at low content occurs as iron (alumino) silicate particles, and at high content on its own.

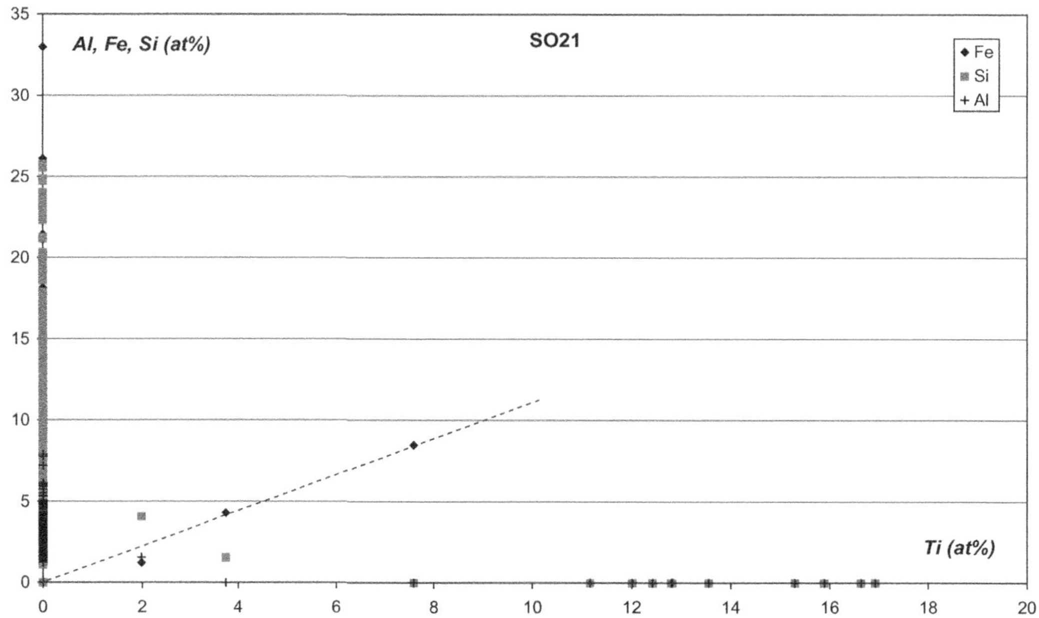

Figure P4.3 Iron, silicon and aluminum as a function of titanium for SO21

Table A4.1 Elemental frequency distribution of all particles

site	total	Al	Ca	Cl	Fe	K	Mg	Mn	Na	P	S	Si
NB50B	1473	18.4	48.9	1.1	27.0	4.8	2.7	0.3	1.7	1.4	3.7	32.3
NB52A	1500	19.6	39.0	11.9	29.7	5.0	5.0	0.3	13.0	0.3	5.5	42.7
RH04	315	21.3	33.0	4.8	31.4	4.4	14.6	0*	7.3	0.3	8.6	48.3
RH06	874	4.6	69.8	11.0	13.8	1.5	1.0	0.1	11.8	0.5	8.5	12.0
SO21	1500	17.6	9.9	0.2	40.9	7.6	2.1	0.3	0.7	0.9	36.5	35.1
SO22	1476	27.0	3.4	0.1	48.0	10.0	1.2	0.4	0.9	0.6	15.7	54.6
TW22	1500	3.2	84.6	8.3	4.1	0.6	1.1	0*	8.1	0.1	3.7	6.6
TW33	1500	11.4	46.6	1.5	19.0	4.5	9.5	0.2	2.2	0.2	5.5	41.5
VE14	1119	4.6	35.7	54.2	7.6	2.1	4.3	0.1	59.3	0*	15.6	11.2
VE22	920	3.8	52.7	41.5	12.2	1.7	5.7	0*	42.9	0.3	2.5	11.8

*absent

This table shows the total number of all particles analyzed in abstracted groundwater from each well and the percentage of particles containing the denoted element. In 4 samples much less than 1500 particles were analyzed.

Obviously, on these membranes the particles were so widely dispersed that the time of the SEM looking for particles became limiting. The time slot of 5 hours, available for analysis, was over before 1500 particles had been analyzed. This exceeding in time confirms the safe sampling time (sampling volume) applied in order to prevent stacking (see Sampling).

Comparison of this table with Table 4.4 shows that in particular iron and sulfur are less represented in particles >2 μm, and all other elements comparable or more.

Chapter 5

Cause and prevention of well bore clogging by particles*

C.G.E.M. van Beek[1], R.J.M. Breedveld[2], M. Juhász-Holterman[3], A. Oosterhof[4] and P.J. Stuyfzand[1,5]

[1]*KWR, Watercycle Research Institute (formerly Kiwa Water Research), P.O. Box 1072, 3430 BB Nieuwegein, the Netherlands; +31 30 60 511; fax +31 30 60 61 165; Kees.van.Beek@kwrwater.nl*

[2]*Water utility Vitens Midden Nederland*

[3]*Water utility Water Company Limburg*

[4]*Water utility Vitens Friesland*

[5]*VU University, Amsterdam*

ABSTRACT

Well bore clogging is caused by accumulation of particles on the well bore. However, all abstracted groundwaters contain particles, but not all wells clog. Apparently, clogging is not caused by the mere presence of particles, but other factors are also involved. Results from on-line well head readings demonstrate that the clogging rate is not constant: the specific capacity decreases during long abstraction periods and remains constant during intermittent abstraction. Results of particle counts reveal that the particle concentration in the abstracted groundwater shows a peak immediately after switching on the well. This concentration peak is caused by the mobilization of particles from the well bore that accumulated during the preceding abstraction period. As long as all accumulated particles are mobilized and removed, no clogging will occur; if not, clogging will occur. Consequently, well bore clogging can be prevented by switching the pump on and off with sufficient frequency to remove the particles accumulated on the well bore during the preceding abstraction period. Moreover, this explanation leads to additional recommendations for well operation, well construction and well design to prevent well bore clogging.

*Published slightly modified as: C.G.E.M. van Beek, R.J.M. Breedveld, M. Juhász-Holterman, A. Oosterhof and P.J. Stuyfzand (2009): Cause and prevention of well bore clogging by particles. Hydrogeol. J. 17 (8) 1877-1886, doi: 10.1007/s10040-009-0537-9

5.1 INTRODUCTION

Well clogging is defined as a decrease in the specific capacity (=volume flow, Q in m^3/h, over drawdown, Δs in m) over time. In well clogging, two types may be distinguished: screen slot (or chemical, inclusive biological) clogging and well bore (or mechanical) clogging. In screen slot clogging, the slots of the screen are clogged, usually by voluminous masses of iron hydroxides, manganese oxides and/or slimes. This type of clogging is easily recognized and has been extensively described in literature (Chapter 7; Houben and Treskatis 2007; Driscoll 1986). In well bore clogging, the gravel pack-aquifer interface and adjacent aquifer are clogged. Because the clogging occurs outside the well, the insides of the wells show no visible signs of clogging, which makes identification confusing. For that reason well bore clogging occurs on a much wider scale than is generally realized.

Up till now, in the Netherlands well bore clogging has only been observed in wells abstracting anaerobic (sulfate-reducing) groundwater from confined unconsolidated sandy sedimentary aquifers. This type of clogging occurs on about one third of all Dutch well fields, but is also reported abroad, in particular in deltaic regions (see Chapter 7). As in the Netherlands sulfate-reducing conditions are present at some depth below groundwater level, usually deep (100–250 meters below surface, mbs), and consequently expensive wells are involved.

By using particle counts, van Beek et al. (1999, 1998) were able to attribute this type of clogging to an accumulation of particles on the well bore. Roscoe Moss Co (1990) also mentioned the role of particles in well clogging, and Blackwell et al. (1995a, b) discussed the role of mica particles, but were not able to support their suppositions with results of measurements.

By taking 10–15 cm long horizontal soil samples over the well bore and adjacent aquifer, followed by micro-morphological research, Timmer et al. (2003) proved that well bore clogging on well field Rodenhuijs (see Figure 5.1 for location of all well fields involved in this research) was caused by the accumulation of particles with a diameter between 2 and 10 μm on the well bore, thereby blocking pore throats. This accumulation develops into a layer on the well bore, which extends no more than 10 cm into the aquifer. This micro-morphological research, confirmed by SEM-EDXRA (scanning electron microscopy – energy dispersive X-ray analysis), also showed that these particles consist of organic matter, calcite and clay minerals. De Zwart (2007) repeated this research on well field Noordbergum and came to the same conclusion. However, the composition of the particles was different, and consisted predominantly of limestone, no organic material and some calcite and clay minerals. This difference in chemical composition of the particles reflects differences in aquifer mineralogy. Micro-morphological research also showed that clogging in Noordbergum was promoted by the presence of remnants of drilling mud on the well bore and in the damage zone. In this contribution, information on the presence and behavior of particles in abstracted groundwater is presented. Based on this information, the mechanism leading to well bore clogging will be explained, along with ways to prevent this type of clogging.

5.2 WELL CONSTRUCTION AND WELL OPERATION

Since 1970, wells in the Netherlands are generally drilled by reverse rotary-air lift, and before 1970 by bailing. After installation of screen and casing, completely of PVC, the wells are developed, usually by a combination of "juttering" (closing the well, pressing the water table down with the help of compressed air, and quickly releasing the built up pressure) and/or pumping with twice the design discharge rate, until the abstracted groundwater is clear and sand free.

All wells are equipped with submersible pumps. Depending on the size of the well, the capacity of these pumps ranges between 60 and 100 m^3/h. Because maintenance on pumps is costly, these pumps are kept in operation until malfunction, after which they are replaced. The number of operation hours/day varies, depending on capacity of well field and drinking water treatment, size of drinking water reservoir and demand.

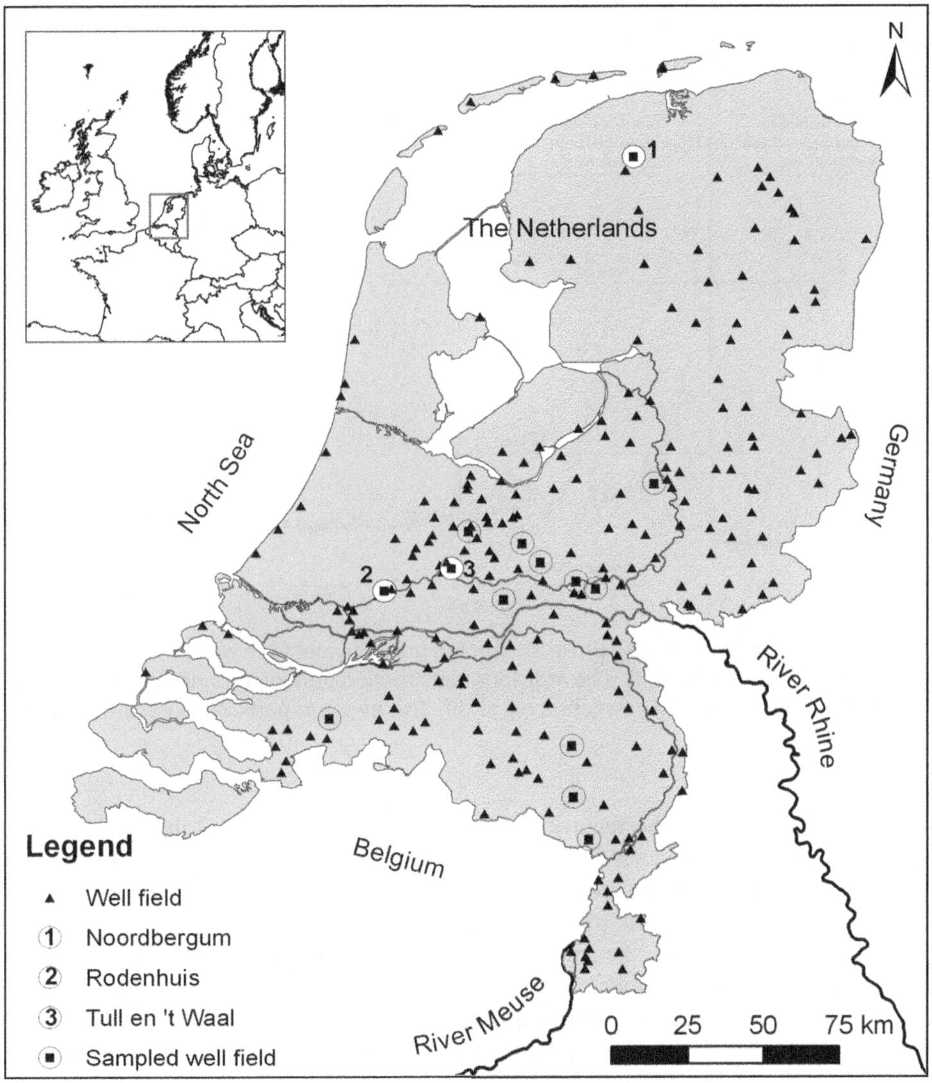

Figure 5.1 Location of all sites mentioned in the text and of all sites sampled

5.3 PARTICLES IN GROUNDWATER

The concentration of particles >2 µm was measured on-line with a particle counter with a frequency of 1 count/minute. The counter was connected to the sample valve on top of the well. Figure 5.2 shows the results of particle counts in abstracted groundwater from well 22 at well field Tull en 't Waal (water utility Vitens Midden Nederland) during a half day. The particle concentration ranges from about 5 to more than 100/ml; in other wells maximum particle concentrations of more than 2000/ml have been encountered.

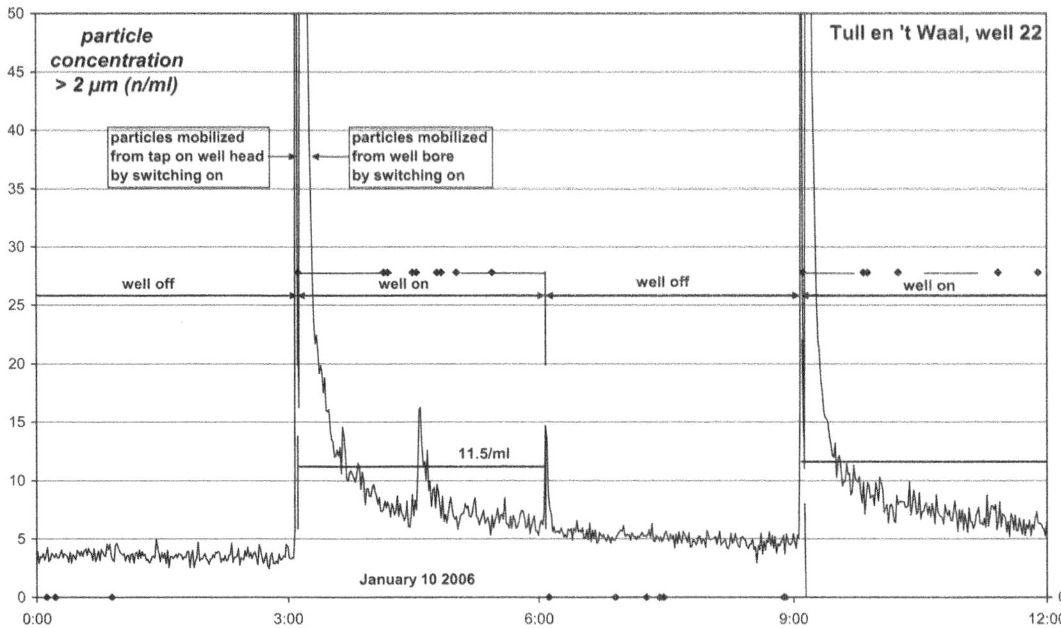

Figure 5.2 Concentration of particles >2 µm in abstracted groundwater from well 22 of well field Tull en't Waal (water utility Vitens Midden Nederland). The individual abstraction periods are indicated by a dotted line, diamonds represent other wells being switched on or off. The average particle concentration during abstraction equals 11.5/ml

Most notable in Figure 5.1 is the recurring, identical shape over time of the particle concentration, characterized by a high peak in concentration immediately after switching on the submersible pump, and quickly leveling off to a more or less constant low concentration. The results of the measurements are reproducible and representative of a phenomenon that occurs each time the submersible pump is switched on. Actually, switching on the submersible pump is characterized by two consecutive peaks in concentration. The first peak coincides with the moment the submersible pump is switched on, and consists of material detached from the sample valve on top of the well. The second peak shows a delay of 5 minutes, equal to V/Q, the volume of water standing in well and gravel pack ($V \approx 10\ m^3$) divided by the discharge rate of the submersible pump ($Q \approx 120\ m^3/h$). During abstraction the particle concentration gradually levels off to an equilibrium concentration of about 6.7/ml. Small peaks are accidental and may be caused by starting submersible pumps in surrounding wells or variations in discharge rate of the pump.

Switching off the submersible pump is occasionally characterized by a single concentration peak followed by a nearly constant concentration. After the submersible pump is switched off, the bottom valve in the pump closes. Since other wells remain in operation, the collector line remains under pressure, and water present in this line flows back through the sample valve and particle counter.

Because this discharge rate (0.006 m^3/h) is small compared to the volume in this line, the particle concentration remains constant.

5.4 PARTICLES IN ABSTRACTED GROUNDWATER

The concentration of particles >2 μm has been measured in groundwater abstracted from several clogging and non clogging well fields, using several wells at each field. All measurements were carried out under normal operating conditions. Clogging is here loosely defined: non clogging wells have never been rehabilitated, and clogging wells one or several times. The results of these measurements are summarized in cumulative frequency diagrams, see Figure 5.3.

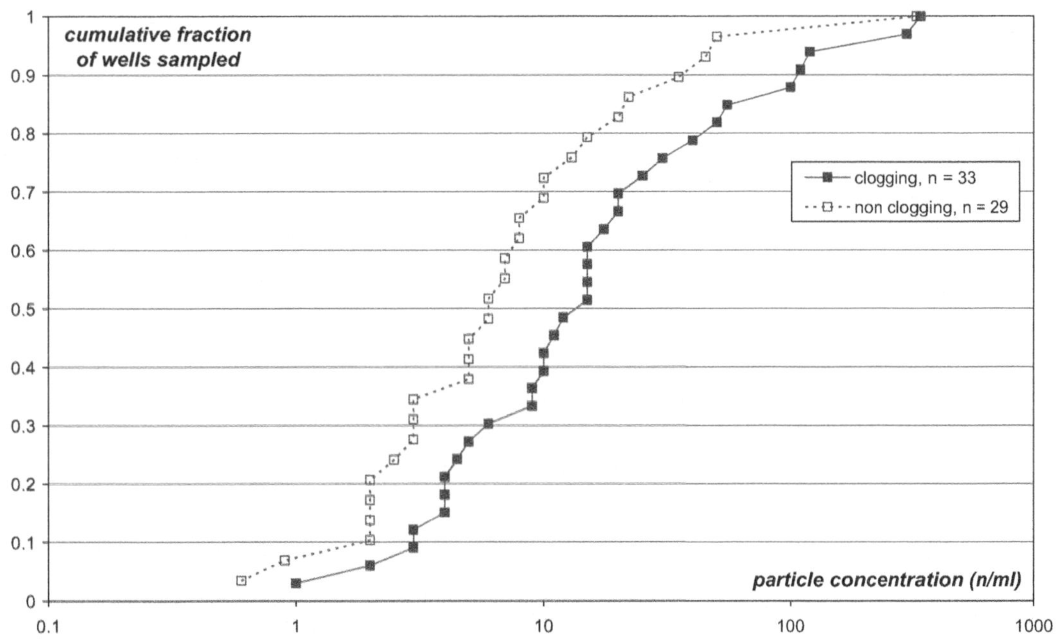

Figure 5.3 Cumulative distribution of the concentration of particles >2 μm in abstracted groundwater under normal operating conditions on clogging (33 wells on 6 well fields) and non-clogging well fields (29 wells on 7 well fields). Each point represents a well

The data points in Figure 5.3 represent the nearly constant concentrations during lengthy abstraction. For example, the particle concentration in the abstracted groundwater from well 22 at Tull en 't Waal (Figure 5.2) is estimated at 6.7/ml. From this figure, it is clear that the concentration of particles with a diameter >2 μm in the abstracted groundwater in 90% of all cases ranges between 2 to 120/ml, with extremes from 0.6 to 700/ml. The cumulative distribution diagram shows that the median concentration in clogging wells equals 15/ml and in non-clogging wells 6/ml. Applying the Mann-Whitney (statistical) test to the data in Figure 5.3 shows that the particle concentration in groundwater from clogging wells is higher (statistically significant, two sided $p < 0.05$) than the concentration from non clogging wells. However, at comparable particle concentrations some wells clog and other wells do not.

In the preceding sections, it was derived from literature that well bore clogging is caused by an accumulation of particles on the gravel pack-aquifer interface and surrounding aquifer. Moreover, it was demonstrated that groundwater, abstracted from clogging and non clogging wells, contains particles. Apparently the presence of particles is a prerequisite, but other processes determine whether well bore clogging will occur or not.

5.5 OCCURRENCE OF WELL BORE CLOGGING

By using on-line pressure transducers, it is possible to measure very frequently the height of the water column standing above the transducer. These transducers have been used on well fields with severe well bore clogging problems to monitor the progress of clogging and to allow timely intervention. As an example, Figure 5.4 shows the height of the standing water column over time in well 50A of well field Noordbergum (water utility Vitens Friesland) from May 2002 till March 2003.

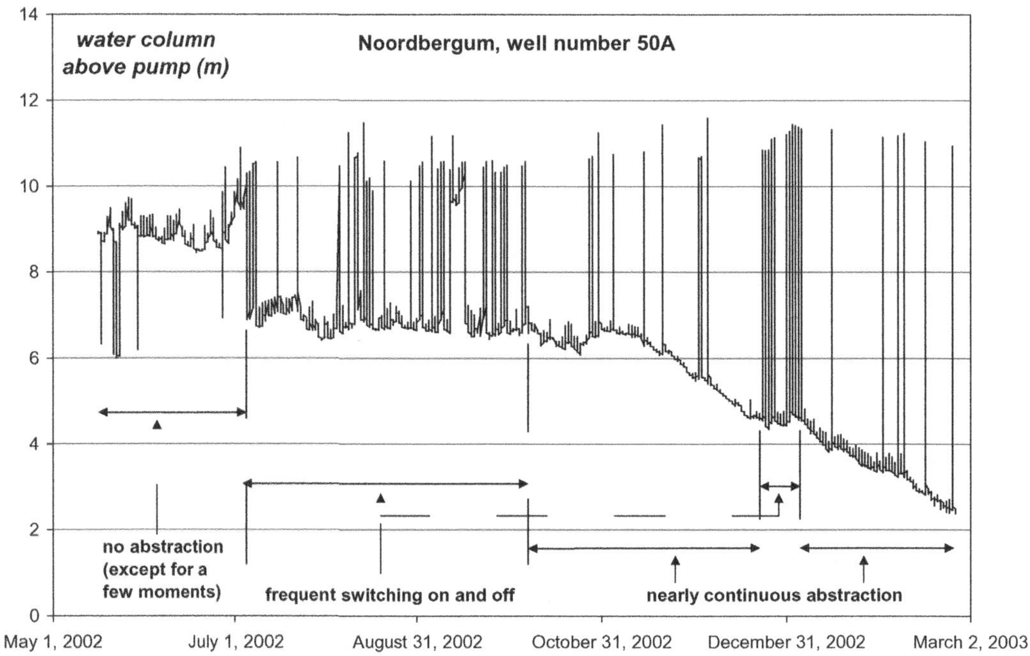

Figure 5.4 Height of the water column over time above the submersible pump ($Q = 65m^3/h$) in well 50A of well field Noordbergum (water utility Vitens Friesland). Before 20 July 2002, the well was out of service, and the water level in the well was influenced by abstraction from neighboring wells

The water level was recorded every three hours. The upper values represent the situation without abstraction, comparable with a high standing water column above the transducer, and the lower values represent the situation during abstraction. Subtraction of the upper and lower values yields the drawdown of the well. Because all recorded heights have been connected by a line, each time the pump is switched on or off appears in this way as a vertical line.

Figure 5.4 shows several notable phenomena. However supply of particles, carried along by the abstracted groundwater, is a continuous process, it is not paralleled by a continuous increase in drawdown. Figure 5.3 clearly shows that the progress in well bore clogging is discontinuous: the drawdown remains about constant from June till October 2002, and increases from October 2002 onwards.

Clogging is governed by switching the submersible pump on and off. In Figure 5.4, the density of the vertical lines indicates the frequency of switching the submersible pump on and off. Closer examination of Figure 5.4 shows that the drawdown remains about constant during periods of frequent switching (July till October 2002 and December 2002/January 2003) and increases during periods of reduced switching or continuous abstraction (November till December 2002 and January till February 2003). This relationship between prolonged abstraction and well clogging was also observed by Juhász et al. (2004). Hoogendoorn et al. (2005) presented an excellent example: on a non-clogging well field, where even drawdown was not monitored, well operation was changed from intermittent abstraction – for drinking water treatment by aeration and filtration – to continuous abstraction – for treatment by membrane filtration. Within a couple of months, all wells were clogged and decommissioned.

Well bore clogging is a fast process. From Figure 5.4, it appears that the drawdown increased 4 meters between 1 November 2002 and 1 March 2003, at a rate of 1 m/month.

5.6 CLOGGING OR NON-CLOGGING?

From the preceding section, it is clear that switching a well on and off influences the rate of well bore clogging. Apparently something happens while switching on a well. Groundwater contains particles (cf. Figure 5.3) that do not all pass the well bore during abstraction. The retained particles accumulate on the well bore (de Zwart 2007; Timmer et al. 2003), and this zone of accumulation slowly expands outwards.

Due to this accumulation, pore-throats on the well bore become constricted and will hinder the flow of water towards the well, resulting in an increase in drawdown. During abstraction the accumulated particles remain "fixed" on the well bore while water keeps flowing towards the well, accumulating still more particles. As soon as the pump is switched off, there is no longer any pressure exerted from water flowing towards the well. Particles accumulated on the well bore may become detached, and particle bridges may collapse. The collapse of particle bridges could be increased by reversing the flow direction by recharge during rest. As soon as the pump is switched on again, these detached particles are carried away by the abstracted groundwater. Moreover, switching on the submersible pump brings water surrounding the well into acceleration. This acceleration results into a larger shear (or erosive) force on the soil matrix, remobilizing retained particles. Consequently, the peaks in particle concentration after switching on the pump (cf. Figure 5.2) are caused by the removal of particles, accumulated on the well bore during the preceding abstraction period.

No clogging will occur as long as, on average, particles accumulated during one abstraction period are removed by the next switching on of the well (van Beek et al. 2004). If not all accumulated particles are removed, clogging will inevitably occur. This reasoning is explained schematically in Figure 5.5 based on particle concentrations in general. The reasoning remains the same for concentrations of particles exceeding a diameter other than 2 μm or for particle mass concentrations (TSS: total suspended solids).

In Figure 5.4, the constant concentration represents the background concentration of particles in attracted groundwater before passing the well bore. This figure also shows the measured particle concentration in the abstracted groundwater. During abstraction, the concentration in the abstracted groundwater is lower than in the attracted groundwater just before entering the well bore; this difference in concentration is retained on the well bore. The areas labeled A, C and E represent a measure of this retained number.

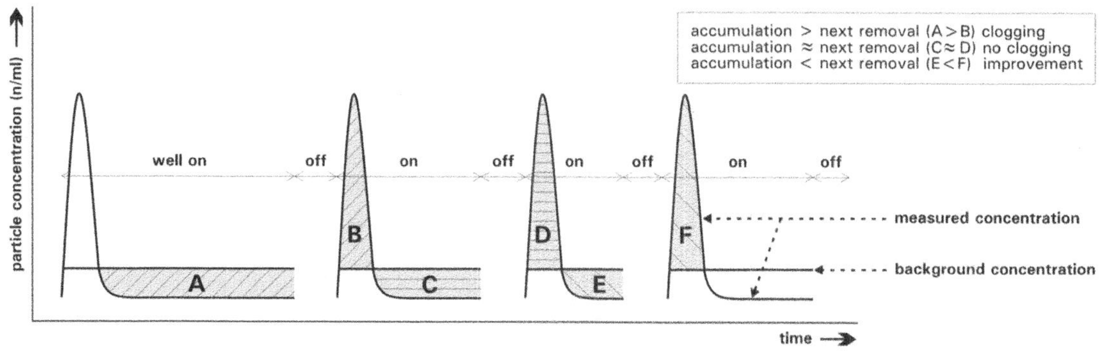

Figure 5.5 Schematic diagram illustrating prevention of well bore clogging

Immediately after switching on the submersible pump, the measured concentration is higher than the background concentration, i.e. more particles are being removed than supplied by the attracted water. The areas labeled B, D and F represent a measure of this removed number. Clogging will not occur as long as, on average, D = C, but will occur if B < A. The well will improve if F > E. This last situation may exist for some time right after drilling, if remnants of drilling mud are removed.

5.7 PARTICLE BALANCE

A particle balance can be determined if the specific capacity of the well remains constant, which means that on average no particles are accumulating on the well bore.

5.7.1 Particles retained on well bore

The number of particles retained on the well bore (and in the sample valve) during an abstraction period equals (Figure 5.5):

$$B_{wb} = Q * (c_{aquifer} - c_{wellhead}) * \Delta t \qquad (5.1)$$

with:

B_{wb}: Number of particles retained on the well bore (n)
Q: Abstraction rate of the submersible pump (m^3/h)
$c_{aquifer}$: Background particle concentration in groundwater in the aquifer (n/m^3)
$c_{wellhead}$: "Plateau" (stationary) particle concentration in abstracted groundwater (n/m^3)
Δt: Length of abstraction period (h)

All particles in the abstracted groundwater originate from the aquifer. Consequently, the particle background concentration can be calculated as the average of the entire concentration graph, including peaks, during one abstraction period by adding up all concentrations and dividing the sum by the number of observations. This calculation yielded an average aquifer concentration of particles >2 μm of well 22 of 11.5/ml, see Figure 5.2. This value is a slight underestimation; the peak of particles after switching off, if present, is not accounted for in this way.

The final observations of long abstraction periods show a nearly constant concentration, which is representative for equilibrium-retaining conditions. This equilibrium concentration varies somewhat within and between individual abstraction periods. The average lowest concentration of particles in abstracted groundwater for well 22 is estimated at 6.68/ml. This means that in this well $(11.50–6.68)/11.5 * 100\% = 42\%$ of the particles carried by the abstracted groundwater are retained on the well bore. This value is also found for other wells.

As the time length of the average abstraction period is available, it is also possible to calculate the number of particles retained on the well bore during an average abstraction period. For well 22 this number equals $1903 * 10^6$.

5.7.2 Particles mobilized from the well bore

The number of particles mobilized has been calculated as the integral of the particle concentration over time minus the "plateau" concentration. Due to the high particle concentration in the peak after the well is switched on, the results of these calculations are not very sensitive to variations in particle concentration and in length of remobilization time.

The first peak, immediately after the well is switched on, consists of particles mobilized in the sample valve on top of the well. For well 22 this was calculated at $425 * 10^6$ particles. The second peak, some minutes after the well is switched on, consists of particles mobilized from the well bore, and was calculated at $1754 * 10^6$ particles. The third peak, immediately after the well is switched off, consists of particles mobilized from the sample valve, and was calculated at $108 * 10^6$ particles.

The particle balance for well 22 is as follows: retained number of particles: $1903 * 10^6 \approx$ mobilized number of particles: $(425 + 1754 + 108) = 2287 * 10^6$. Both numbers compare reasonably well, and this similarity in number of accumulated and removed particles is considered as evidence that well bore clogging is indeed governed by a particle balance. These calculations have been executed for 4 wells, two old wells (numbers 21 and 22) and two new wells (numbers 33 and 35), and the differences in the number of particles mobilized are surprising: $43 * 10^6$ to $2287 * 10^6$ particles.

The number of particles mobilized is determined in particular by the acceleration force exerted on the particles, i.e. the capacity of the pump and the area of the well bore, and by the number of particles accumulated. Additional factors are the distribution of the accumulated particles over the length of the well bore, the distance to the well screen and the accumulation circumstances (for example, interaction with remnants of the drilling mud), but these factors are currently intangible. Because the flux (discharge rate per unit area) over the (open) well bore of well 21 and 22, and of well 33 and 35, are comparable, the number of mobilized particles appears to be limited by the number of accumulated particles.

As the individual abstraction period increases, more particles are accumulated on the well bore, and more particles will be mobilized by the subsequent start of the pump. To investigate this assumption, the number of particles removed from the well bore is presented in Figure 5.6 as a function of the length of the preceding abstraction period. This figure shows no clear relation between the number of particles removed and length of the preceding abstraction period. However, two qualitative remarks seem possible: lower numbers of removed particles are limited to short preceding abstraction periods, in particular in well 33 and 35, and apart from the outlier in well 22, there seems to be a maximum in the number of removed particles for each well. Moreover, since the wells are non-clogging, the number of particles removed must be equal on average to the number retained.

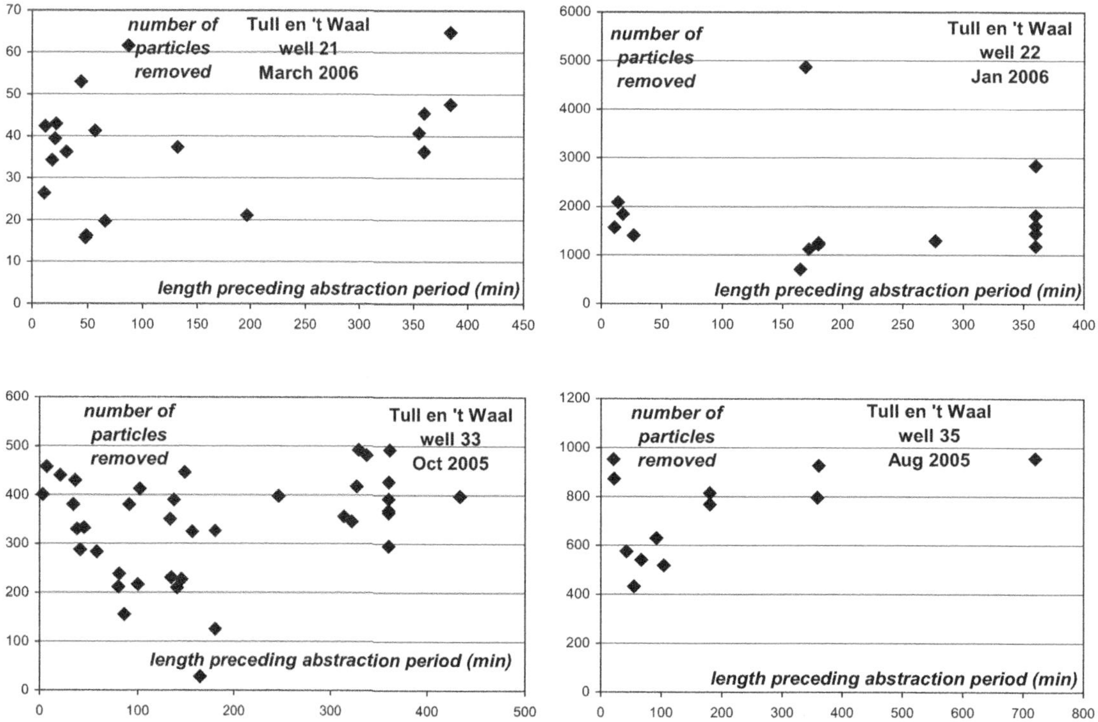

Figure 5.6 Number of particles removed ($* 10^6$) after starting the pump in four wells, as a function of the length of the preceding abstraction period

Notable are the large differences in the number of particles mobilized from the well bore after switching on the submersible pump: from $15 * 10^6$ to $65 * 10^6$ in well 21 to $700 * 10^6$ to $2800 * 10^6$ in well 22 ($4800 * 10^6$ is considered an outlier). Because, at well field Tull en 't Waal, well constructions and pump capacities are comparable, the number of removed particles must be governed by the particle supply of the aquifer. This makes well 22 most, and well 21 least vulnerable to well bore clogging.

5.8 RECOMMENDATIONS FOR WELL OPERATION, CONSTRUCTION AND DESIGN

The preceding sections, with the help of a particle balance, have confirmed that well bore clogging is caused by accumulation of particles on the well bore, and that this type of clogging can be prevented by regularly switching the well on and off to remove these accumulated particles. However, other possibilities for preventing well bore clogging can be deduced from the specific particle accumulation on the open well bore:

$$B_{area,owb} = \frac{1}{f_Q} * B_{spec,wb} * \Delta t = \frac{Q}{\pi * D * L} * \frac{1}{f_Q} * (c_{aquifer} - c_{wellhead}) * \Delta t \quad (5.2)$$

with:

$B_{area,owb}$: Specific particle accumulation on open well bore (n m^{-2})
f_Q: Actual specific capacity/initial specific capacity (-)
$B_{spec,wb}$: Specific particle accumulation rate on well bore (n m^{-2} h^{-1})
D: Diameter of the well bore (m)
L: Length of the well screen (m)
Δt: Individual abstraction period (h)

Equation (5.2) clearly shows that the accumulation of particles on the well bore may be reduced, and consequently the risk of well bore clogging, by decreasing the abstraction rate Q, by increasing the diameter of the well bore D and the length of the well screen L, by decreasing the number of particles retained on the well bore ($c_{aquifer} - c_{wellhead}$) and by decreasing the length of the individual abstraction periods Δt. This leads to additional recommendations for well operation, well construction and well design. Moreover, from this equation, it may be deduced that the accumulation rate is inversely related to the degree of clogging: the more a well gets clogged, the faster the process. This experience has been encountered in the field by many well operators.

Conversely, for non-clogging conditions, the equation shows that the discharge rate and/or the individual abstraction period can be increased. Increasing the discharge capacity of non-clogging wells leads to substantial savings.

5.8.1 Well operation

Reduction in discharge rate (Q) results in a decreased particle load. Besides this direct effect, there is also an indirect effect: a reduction in discharge rate leads to a reduction of groundwater flow velocity and consequently to a reduction in particle concentration (caquifer), especially of larger particles (Chapter 3). The particle concentration is in turn governed by the particle pool in the aquifer, but this pool is beyond control.

Finally, Equation (5.2) clearly shows that well bore clogging may also be prevented by shortening the individual abstraction periods. This was confirmed experimentally (Chapter 6).

5.8.2 Well construction

Probably the most efficient way to prevent well bore clogging is to minimize the retaining properties of the well bore, i.e. minimizing ($c_{aquifer} - c_{wellhead}$). If there are discontinuities in grain size, small grains will partially fill up the pores between the large grains, resulting in smaller pore throats than in either of the contributing layers. This will also occur at the interface between the large grains of the gravel pack and the small grains of the aquifer. Therefore, even under theoretical conditions, there is a "dip" in pore throat diameter at the well bore. However, experimental work by Bradford et al. (2005) showed that particles are only retained at discontinuities from coarse to fine.

Drilling mud may be "contaminated" by fine sand originating from fine sandy layers that were drilled through. This fine sand becomes trapped between the aquifer and the gravel pack. During development of the well, only material with a diameter smaller than the pore-throats of the gravel pack can be removed. The finer the gravel of the gravel pack, the finer the sand that cannot pass this pack during development. All material (contamination) that is carried with the drilling mud which has a diameter too large to pass the pore throats of the gravel pack will remain trapped on the well bore. This trapped material will minimize the porosity and diameter of the pore throats. Figure 5.7 illustrates this situation schematically. This minimized diameter of the pore throats, even more than the minimized porosity, enhances accumulation of particles carried by groundwater flowing towards the well.

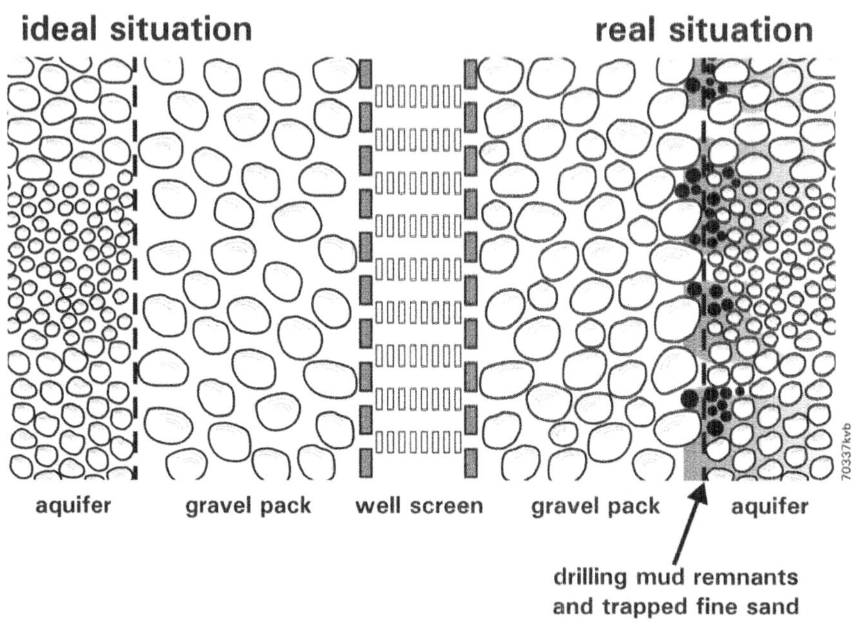

Figure 5.7 Schematic representation of the well bore after traditional development with trapped fine sand grains (black), peat and wood fragments, and remnants of the mud cake (gray shading)

This "dip" in pore throat diameter is aggravated by remnants of drilling mud on the well bore or in the damage zone. During well development, not all the mud may be removed. Development is intended to remove the mud cake applied during drilling, but the criteria to determine whether a well is developed or not are the production of clear water and the absence of sand grains at a discharge rate exceeding the design value (Driscoll 1986; Roscoe Moss Co 1990; DVGW 2002). In fact, well development should continue until the specific capacity of the well no longer increases, indicating that the mud cake has been removed as completely as possible. With the current practice, remnants of mud cake will certainly remain after well development. Evidence of this incomplete development is shown by the specific capacity of clogged wells (van Beek 1984) and of newly commissioned wells (Breedveld et al. 2007) exceeding the value on commissioning after rehabilitation, a spontaneous increase of the specific capacity of newly commissioned wells during operation and by the results of micro-morphological research showing remnants of drilling mud (de Zwart 2007). In column experiments, Deelder (1977) also showed that the mud cake is difficult to remove. After forced abstraction, some small holes were pulled in the mud cake, but the larger part remained intact.

In fact, with coarse aquifers, two favorable conditions coincide: the larger the grain size of the aquifer, the coarser the gravel that can be applied for the gravel pack. A larger diameter of the pore throats allows both better development of the well and easier passage of particles deposited on the well bore during drilling. As a result, compared to coarse textured aquifers, fine-textured aquifers are twice as vulnerable to well bore clogging. This effect is aggravated by the tendency to apply finer gravel packs in order to avoid sand production from the well (Hermann and Tholen 2005).

The minimized diameter of the pore throats and porosity caused by the presence of fine material trapped on the well bore can be counteracted by drilling in various ways. One possibility is to use a de-sander, which completely removes fine sand and floating material. Another possibility is to use removable or degradable material as drilling mud, like calcium carbonate removable with acid, or organic material degradable under all conditions. The well bore can also be scraped just before installing the well screen (Olsthoorn and van Harlingen 1994), which removes invaded drilling mud (Howsam and Hollamby 1990; Dudgeon and Cox 1976). Yet another possibility is to use bailer drilling, where no drilling mud is applied. The success of this recommendation has been confirmed in the field by Timmer et al. (2003). Wells drilled by bailing clog much more slowly than wells drilled by reverse rotary-air lift. This observation can be easily explained. Bailing smears the well bore, but no drilling mud and CMC (drilling additive) mixed with fine sand and organic debris are used. Finally, clogging will be retarded or prevented by removing as much drilling mud as possible, which results in maximum development, instead of simply developing the well until clear and sand-free water is obtained (Breedveld et al. 2007).

5.8.3 Well design

As the surface area of the well bore (πDL) increases, the specific accumulation of particles decreases, and the well becomes less susceptible to clogging. Consequently, enlarging the diameter of the well bore and increasing the length of the well screen are advantageous. When enlarging the diameter of the well bore, the diameter of the well screen should also be increased to prevent the gravel pack from becoming too thick. Obviously, in wells with a thick gravel pack the well bore is harder to reach than in wells with a thin gravel pack, aggravating development and rehabilitation.

In well design, often a screen as long as possible is applied, including intercalated fine-grained layers. However, the diameter of the gravel in the gravel pack will be determined by the finest aquifer layer exposed to the well screen. From the foregoing, it is clear that the finer the gravel pack, the more vulnerable the well is to well bore clogging. For this reason it is advised to limit the well screen to the coarsest parts of the aquifer, with a gravel pack that is as coarse as possible (Breedveld et al. 2007). Another approach consists of applying various gravel packs with gravel diameters adapted to the aquifer grain size over the length of the well screen.

5.9 DISCUSSION AND CONCLUSIONS

Well bore clogging is caused by the accumulation of particles on the well bore. However, even though all sulfate reducing ground waters contain particles, not all wells clog. This phenomenon can be explained by means of a particle balance: as long as on average all particles accumulated on the well bore during each abstraction period are removed by the subsequent switching on of the submersible pump, no clogging will occur. As soon as on average all accumulated particles are not removed by the subsequent switching on of the pump, clogging will inevitably occur. Consequently, to prevent this type of clogging, the wells should be operated intermittently. Note that this recommendation is opposite to that for wells vulnerable to screen slot (or chemical) clogging. Such wells should be abstracted continuously, and the pump should be switched on and off as little as possible (Chapters 7 and 8).

Introducing this well operating method to existing well fields depends on local conditions: degree of clogging of individual wells, number of wells, presence of differences in chemical composition of abstracted groundwater between individual wells, drinking water treatment facility and volume of drinking water reservoir. These conditions may limit application of this operating method, or make adaptations necessary. For instance, the number of wells may have to be increased. Van der Wurf et al. (2007) devised a very elegant operating scheme, with a maximum for each abstraction period and consequently for the number of particles accumulated. Depending on vulnerability for clogging, the wells were divided into several groups. Between the groups, large differences in the number of operation hours may exist. Because within each group the actual abstraction periods for each well shift over the day, the numbers of operation hours within each group are comparable.

In addition to intermittent abstraction, well bore clogging can also be prevented by minimizing the particle-retaining properties of the well bore. This leads to recommendations for well operation, construction and design. Probably because no clogging material is visible, some consider this well deterioration not as clogging, but as a delayed effect of incorrect well construction. Regardless of what it is called, this type of well clogging is widespread in wells abstracting anaerobic (sulfate-reducing) groundwater from unconsolidated sandy sediments.

Chapter 6

Prevention of well bore clogging by intermittent abstraction*

C.G.E.M. van Beek[1], Rob Breedveld[2], Martijn Tas[2] and René Kollen[2]

[1]*KWR Watercycle Research Institute*

[2]*Vitens*

ABSTRACT

Well field 'Tull en 't Waal' (water utility Vitens Midden Nederland) suffers extensively from well bore (or mechanical) clogging. To prevent this type of clogging, wells have been operated intermittently since August 2005 using a carrousel well operation schedule. Since the introduction of this schedule, well bore clogging has become minimal.

6.1 INTRODUCTION

When groundwater is abstracted from unconsolidated sedimentary aquifers, two types of well clogging can be distinguished (van Beek 1984; Houben and Treskatis 2007; Chapter 7): screen slot or chemical clogging and well bore or mechanical clogging. Screen slot clogging occurs if a single well abstracts both oxygen-containing water and iron-containing water. Upon mixing, iron hydroxide precipitates are formed, resulting in well clogging. This process has been described in greater detail elsewhere (Chapter 7 and 8). In the Netherlands this type of clogging is limited to (oxic) freatic aquifers.

In literature, particles have been suggested as a cause of well clogging by van Beek (1989), and Roscoe Moss (1990), Blackwell et al. (1995a, b) specifically addressed mica particles. The presence of particles was confirmed by van Beek (1998, 1999, Chapter 3) in abstracted anoxic (sulfate-reducing) groundwater by using particle counters. Clogging of the well bore by accumulation of particles was made reasonable by van Beek (2002), and was experimentally confirmed by Timmer et al. (2003) and de Zwart (2007) by means of soil micro-morphological research.

*Published slightly modified as: Kees van Beek, Rob Breedveld, Martijn Tas and René Kollen (2010): Prevention of well bore clogging by intermittent abstraction. Ground Water Monitoring & Remediation, doi: 10.1111/j1745-6592.2010.01307.x

During pumping, particles may accumulate on the well bore, resulting in well bore clogging. Since this clogging process occurs at some distance from the well, camera inspection shows these wells to be completely clean inside, which can lead to confusion. These particle accumulations may dissipate during rest and the particles carried away with the abstracted water after switching on the pump again. Consequently, this type of clogging can be prevented by regular alternation of abstraction and rest (Chapter 5). This method of well operation has now been used on the well field of Tull en 't Waal (water utility Vitens Midden Nederland) for more than 4 years.

This paper describes the results of this intermittent abstraction. These results were collected during the normal operation of the public drinking water facility, that is, the operation of the plant was not altered in order to show these effects more clearly. The results are favorable. Previously, each well had to be rehabilitated every 2 to 4 years; during the last 4 years, however, only 1 well out of 12 needed rehabilitation.

6.2 WELLS AND METHODS

6.2.1 Well field and wells

Well field Tull en 't Waal is located in the center of the Netherlands; it abstracts groundwater from unconsolidated sedimentary aquifers at depths between 85 and 165 mbs (meters below surface). The abstracted groundwater is chemically characterized by the occurrence of sulfate reduction, the presence of traces of methane and by calcite saturation. Total annual production during the period 2005 to 2008 varied between 3.3 and 4.3 million m^3/year, which was produced by 12 wells. Due to the extension and renovation of the treatment facility, which required the replacement of 3 wells, the well field consisted temporarily of 15 wells.

Construction details for the wells referred to in this paper are summarized in Table 6.1.

6.2.2 Carrousel abstraction schedule

In order to arrive at a regular alternation of abstraction and rest, a rotating operation or carrousel schedule was developed (van der Wurf et al. 2007). This schedule has been in operation since August 17, 2005 and is shown in Table 6.2.

This schedule has five carrousels in which well operation rotates. All wells operate based on the water demand level in the drinking water reservoir (level 1 represents low demand with a nearly full reservoir, and level 9 the highest demand with a nearly empty reservoir). The demand level in the reservoir over the day usually compares with level 4; level 5 appears occasionally, and levels 6 to 9 only rarely, during situations such as fire fighting.

Within each carrousel, all wells move up to the next demand level every 3 hours: the well that was running at that moment switches off, and moves to the third (last) position. The second well moves to the first position and starts operating. If this well was already operating, it stays in operation, and therefore may operate for 6 hours. The third well moves to the second position and will start operating or not depending on the water level in the drinking water reservoir. As is apparent from the schedule in Table 6.2, wells in carrousels 1 and 5 have the most production hours and will often operate for 6 hours and rest for 3 hours, while the wells in Carrousel 2 have the fewest production hours.

The time interval of 3 hours of the carrousel scheme is based on experiences of the plant operator, trial and error, and theory (Chapter 5). On this moment, this time interval of 3 hours is also applied on other well fields sensitive to well bore clogging.

Table 6.1 Construction details of three representative wells

Well number	25		28	33
Year of construction	1977		1980	2004
Drilling method	Reverse rotary – air lift		Reverse rotary – air lift	Reverse rotary – air lift
Drilling additive	Bentonite		Bentonite	CMC
Bore diameter (mm)	700		800	1000
Final depth of drilling (mbs*)	150		173	126
Well casing (mbs)	0–32.6	32.8–125.3	0–135.0	0–84.4
Diameter (mm)	360/400	285/315	360/400	461/500
Material	PVC	PVC	PVC	PVC
Screen (mbs)	125.3–144.0		135.0–147.0	84.4–89.2
Blank casing (mbs)			147.0–155.0	89.2–93.9
Screen (mbs)			155.0–165.0	93.9–107.0
Blank casing (mbs)				107.0–116.2
Screen (mbs)				116.2–121.0
Screen diameter (mm)	300/305		360/400	461/500
Material	Stainless steel		PVC	PVC
Slot width (mm)	1		1	0.75
Filter pack material size (mm)	1.5/2.5		1.5/2.5	1.0/1.6
Screen total length (m)	18.7		22.0	22.7
Bottom sump (mbs)	144.0–144.5		165.0–166.0	121.0–121.5
Bottom sump diameter (mm)	300/305		360/400	461/500
Material	Stainless steel		PVC	PVC

* mbs: meter below surface

Table 6.2 Scheduling of the wells per carrousel and per level

Carrousel	Well number	Water Demand Levels Drinking Water Reservoir								
		1	2	3	4	5	6	7	8	9
1*	20, 21, 22	X			X			X		
2	24, 25, 26					X	X			X
3	23, 27, 28	X					X			X
4	29, 30, 31			X				X		X
5	33, 34, 35		X		X				X	

* decommissioned in October 2006

Each carrousel consists of three wells that shift one position every 3 hours, which means that the carrousel is back at its original position after 9 hours. Since 24 hours is not divisible by 9 hours, the composition of the active wells also shifts over the day. The initial composition is reached again after 8 days. This shift over the day results in comparable running hours for all wells in the same carrousel.

After the introduction of this carrousel schedule, many changes were made. For instance, carrousel 1 was closed due to the abandonment of the three wells (20, 21 and 22). Additionally, wells have been shifted regularly within their carrousel. As the plant operators want the drinking water reservoir filled, an upward shift (a shift to the left in Table 6.2) results in a more frequent operation of the corresponding carrousel and thus in an increase in the number of operating hours. A downward shift (a shift to the right in Table 6.2) results in a decrease. Since the total production (which is approximately equivalent to the total number of operation hours) remains the same, shifting a well within a carrousel also affects the number of operation hours of wells in other carrousels. The magnitude of these changes may vary between a considerable change (a shift of several wells over several levels) to a negligible one (a shift over one level in another carrousel). The distribution of the wells within the carrousels has not been changed.

A SCADA (Supervisory Control And Data Acquisition) system records when the wells are on and off. From these records several operational variables may be calculated, such as the time lengths of the consecutive operation and rest periods, the number of operation hours/months, the number of times the wells have been switched on and off, and so on.

6.2.3 Well head measurements

Water level pressure transducers were installed at all the wells of well field Tull en 't Waal. The diameter of the monitor pipe inside the well was too small for internal installation, so the pressure transducers were installed in the piezometers constructed in the gravel pack of each well. These transducers recorded the pressure head of water every 30 minutes.

Two levels of pressure head are apparent from the data (Figure 6.1), an upper level that represents the well condition during rest, and a lower level that represents the well condition during abstraction.

Since the atmospheric pressure was not always available, the difference between the two pressure head levels was used for the drawdown instead of the difference between the two levels of hydraulic head. This approximation is acceptable, because we are not as interested in an accurate value for the drawdown as we are in the changes over time. Drawdown was estimated as the difference between two consecutive readings, provided this difference exceeded 1 m:

$$\text{If } (h_{t=1} - h_{t=2}) > 100\,\text{cm}, \text{ then } \Delta s = p_{t=1} - p_{t=2} = h_{t=1} - h_{t=2} + a_{t=1} - a_{t=2}.$$

where Δs is drawdown, p is hydraulic head, h is pressure head, a is atmospheric pressure (all in cm H_2O), and $t = 1$ and $t = 2$ are the times of two consecutive readings, with a 30-minute interval.

Changes in atmospheric pressure over 30 minutes are generally confined to less than 3 mbar ($\approx$0.3 cm H_2O), resulting in a negligible contribution of ($a_{t=1} - a_{t=2}$) to drawdown. Consequently, in actual practice, the drawdown equals the difference in pressure head.

Both levels in Figure 6.1 show some variation over time. This variation is caused by whether or not neighboring wells were abstracting (mutual influence). The lower level shows some additional variation due to variations in discharge rate (the discharge rate abstracted by a well fluctuates with the number of wells delivering simultaneously on the same collector line: the rate is higher if the well is operating alone and lower if more wells are delivering on the same line). Moreover, Figure 6.1 shows that even if the well is not operating, the variation in the upper level remains the same. Obviously, in a practical sense, equilibrium in water level was reached within 30 minutes. This is confirmed by the near absence of data points between both levels.

6.3 RESULTS

As already mentioned, the operation of the wells within each carrousel is comparable, but the wells may differ in construction (see Table 6.1), drilling history and development, clogging history, particle concentration in abstracted groundwater and capacity of the pump.

Carrousel 3, comprising wells 23, 27 and 28

Figure 6.1 shows the relationship between recorded pressure head and calendar date for well 28, where the upper water level represents the conditions during rest and the lower water level the conditions during abstraction. Simultaneous absence of data points in the upper and lower water levels represents data that is lacking. Absence of data points in the upper level, together with data in the lower level, indicates prolonged periods of abstraction. Conversely, absence of data point in the lower level, together with data in the upper level, indicates rest.

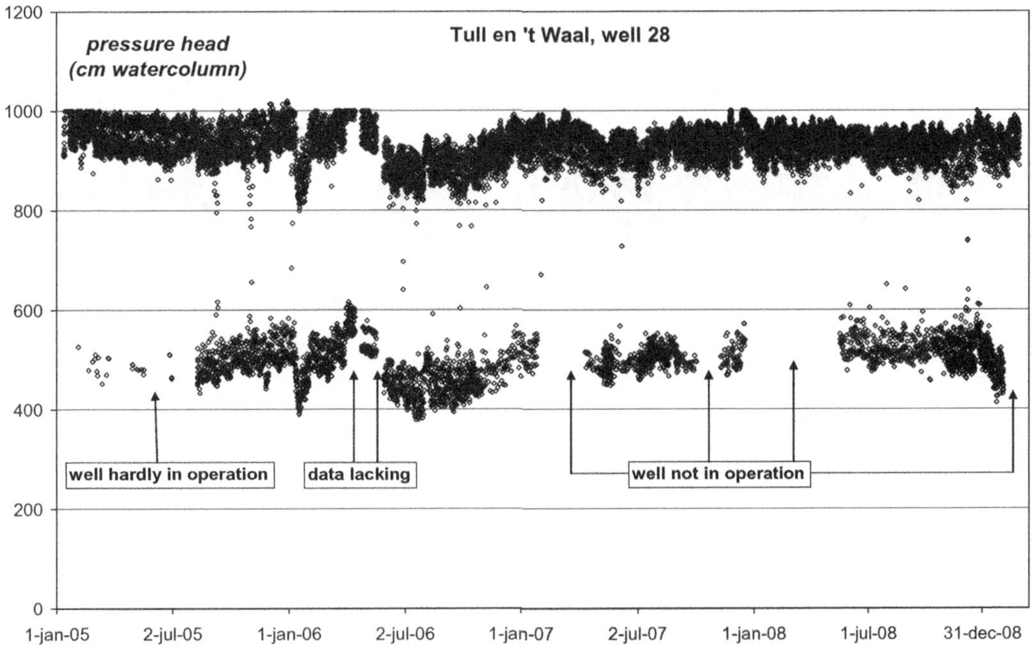

Figure 6.1 Pressure head plotted against calendar date for well 28. Until 24 April 2006, the rest level approached the upper level of the measurement range of the pressure recorder, but this has a negligible effect on the calculated drawdown. Data were lacking between 14 and 26 April, and 17 and 29 May 2006. The well was not in operation between 26 January and 12 April 2007, 3 October and 9 November 2007, 12 December 2007 and 16 May 2008, and after 3 February 2009. Before the introduction of the carrousel system on 17 August 2005, the well was used only occasionally

The difference between the two levels in Figure 6.1 was used to calculate drawdown shown in Figure 6.2. This figure also shows the number of operation hours per month as a derived measure for intermittent abstraction: the fewer the operating hours, the longer the rest periods. Comparison of the number of operation hours in Figure 6.2 with drawdown confirms that the well was idle several times for some months.

Figure 6.2 shows that drawdown decreased in well 28 from the start to 10 April 2007, remained fairly constant until 4 November 2008 and then increased sharply. This picture of drawdown is more or less paralleled by the number of operating hours: a steady improvement with approximately 240 hours/month, but no improvement after operation on 1 November 2006 decreased to approximately 70 hours/month. The idle period from 2 February to 10 April 2007 resulted in some further improvement. Later, notwithstanding the low intensity of abstraction and the long idle periods, there was no further improvement. Obviously, the room for improvement (the extent of decrease of drawdown) is limited by the condition at commissioning. The room for increase in drawdown, on the other hand, is nearly unlimited. Note the abrupt increase in drawdown after 4 November 2008, which coincides with an increase in the number of operation hours to around 360 each month.

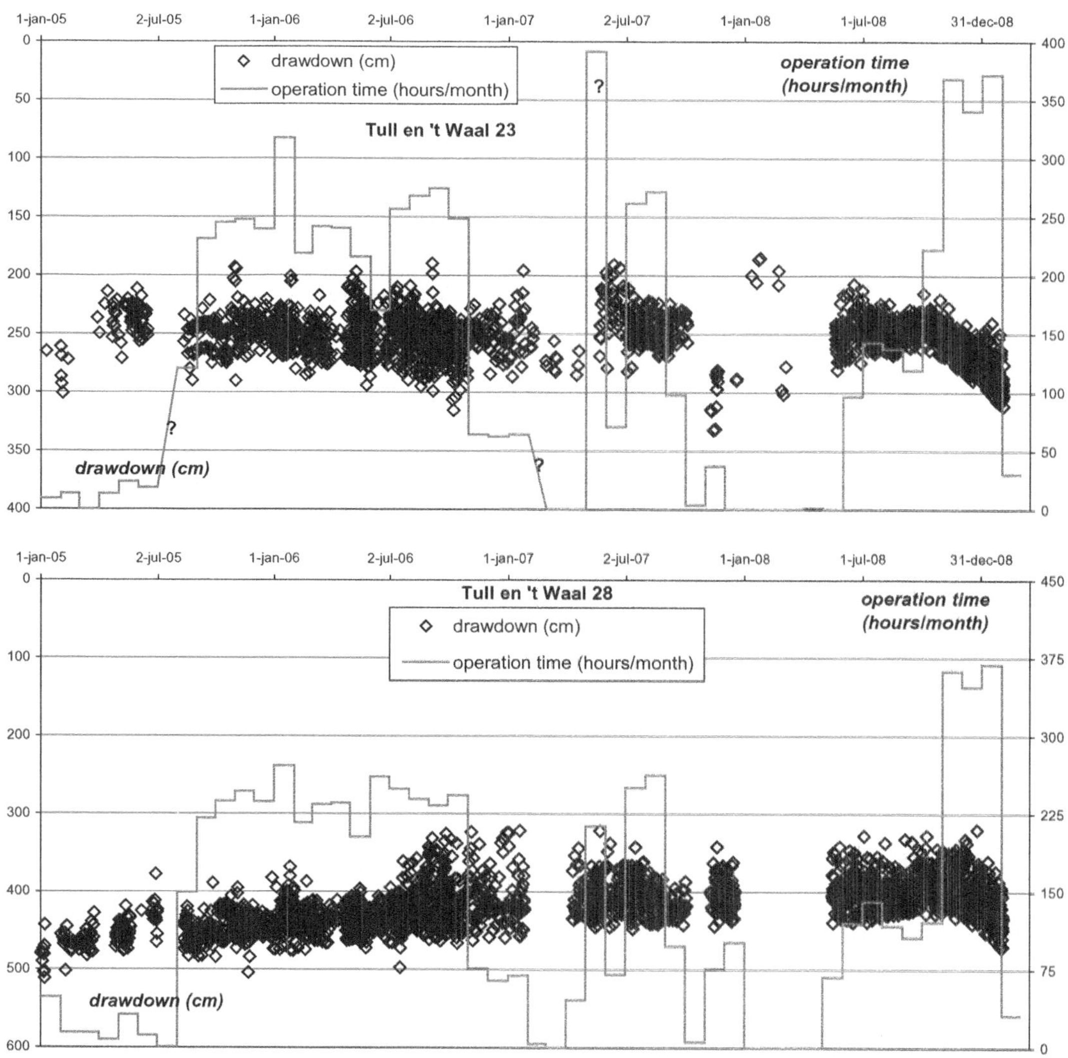

Figure 6.2 Drawdown (left axis) and operation hours/month (right axis) over time for well 23 (upper part) and well 28 (lower part). Well 27 is not shown because the actual pressure exceeded the measurement range

It is clear from the preceding paragraph that an abrupt, adverse change in the number of operation hours/month results in an immediate increase in drawdown and that well 28 can be operated for 240 hours/month without clogging, but operation at 360 hours/month results in clogging.

Figure 6.2 also shows drawdown in well 23 over a period of time. Drawdown remained constant until 4 November 2008 and increased sharply thereafter. Obviously, drawdown could not decrease, as there was hardly room for improvement.

Carrousel 5, comprising wells 33, 34 and 35

Figure 6.3 shows drawdown and operation hours for well 33 as functions of calendar date. On 10 January 2006 the submersible pump with a capacity of about 70 m^3/h was replaced by a larger one with a capacity of about 100 m^3/h. Drawdown on 10 January 2006 measured about 140 cm and on 2 February about 270 cm. Drawdown as a result of the replacement of the small pump by a larger one corresponds with $100/70 * 140 = 200$ cm. The increase in drawdown as a result of continuous abstraction during 3 weeks consequently amounted to $270 - 200 = 70$ cm ($\approx$11.2 m/year), and was caused by well clogging. This occurrence of well clogging was paralleled by a decrease in specific capacity: beforehand $70/1.40 = 50$ m^3/h per m drawdown, and afterwards $100/2.70 = 37$ $m^3/h.m^{-1}$, a decrease of $(50–37)/50 * 100\% = 26\%$.

After 21 April 2006 drawdown continued to slowly increase, but increased sharply after 17 November 2006. On 22 September 2006 and 17 November 2006 the carrousel schedule was modified, resulting in increased operating hours. Drawdown increased after 22 September, but even more so after 17 November. This appeared to be a somewhat delayed reaction to an unfavorable change in operation. Unfortunately, this increase in drawdown escaped attention.

Between 16 June and 20 July 2007 the well was rehabilitated. The specific discharge of the well before rehabilitation amounted to about $100/5.10 = 19.6$ $m^3/h.m^{-1}$, and afterward to about $100/3.40 = 29.4$ $m^3/h.m^{-1}$. Because clogging is located on the well bore, these wells are difficult to rehabilitate. Moreover, clogging of the well exceeded 50% (the actual specific discharge amounted to less than 50% compared with the value on commissioning), which is the utmost limit recommended for rehabilitation of clogged wells (van Beek 1984).

Operation became less intensive after rehabilitation. Drawdown remained fairly constant or increased slowly. Apparently, operation at 240 hours/month is the maximum that well 33 can withstand.

The decrease in drawdown, after the well had been out of operation between 2 November 2008 and 9 February 2009, is in line with the expectation. Note the slow increase in drawdown during July to September 2006 and the nearly constant drawdown during July 2007 to November 2008, with fewer operating hours in the first period than the second period. Apparently, expressing intermittent abstraction of the carrousel in hours/month is too simple; the picture describing drawdown is somewhat more complicated.

Figure 6.3 also shows that drawdown in well 34 increased by about 80 cm between November 2006 and April 2008, with 300 production hours/month, and later on remained constant with 240 hours/month. Well 35, even with a larger submersible pump, showed no clogging at all.

Carrousel 2, comprising wells 24, 25 and 26

Figure 6.4 shows drawdown and operation hours for well 25 as functions of calendar date. This figure shows that the well improved considerably between 17 August 2005 and 8 January 2006 as a result of the negligible abstraction. Drawdown increased immediately after an increase in operation hours on 28 January 2007. This increase ended and drawdown remained constant during the period of low abstraction between 5 July and 5 September 2007. After 5 September drawdown remained constant or even decreased somewhat until 17

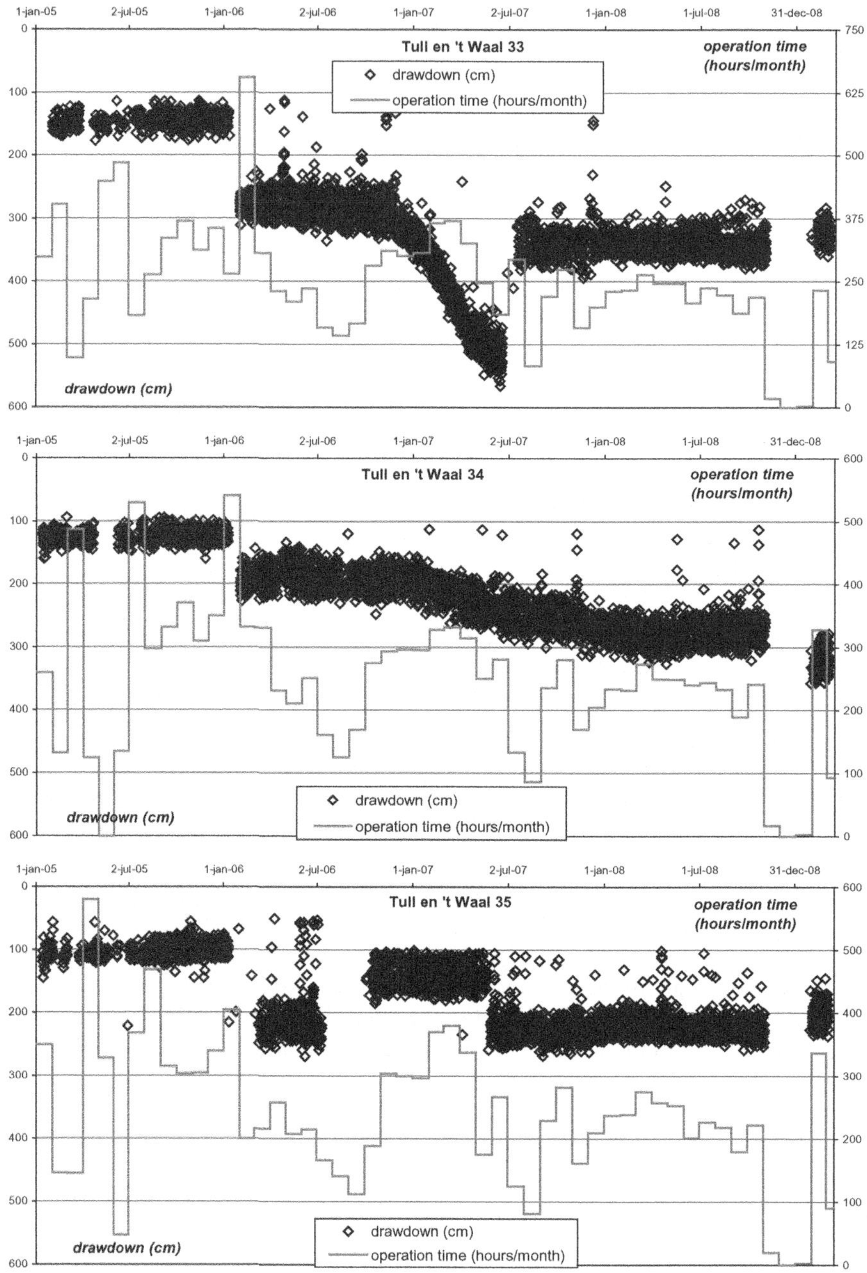

Figure 6.3 Drawdown (left) and operation hours per month (right) as function of calendar date for well 33 (upper part), 34 (middle part) and 35 (lower part). Before introduction of the carrousel schedule on 17 August 2005 the wells were in operation and idle for long periods. Between 10 January and 2 February 2006 wells 33 and 34 abstracted continuously with variable discharge, as did well 35 between 20 February and 3 March 2006. Between 30 June and 17 July 2007 well 33 abstracted intermittently after rehabilitation. Between 3 November 2008 and 28 January 2009 all wells were not in production

December 2007, after which drawdown remained constant or increased very slightly. Note that drawdown started to increase with fewer operating hours than during the period that drawdown remained constant.

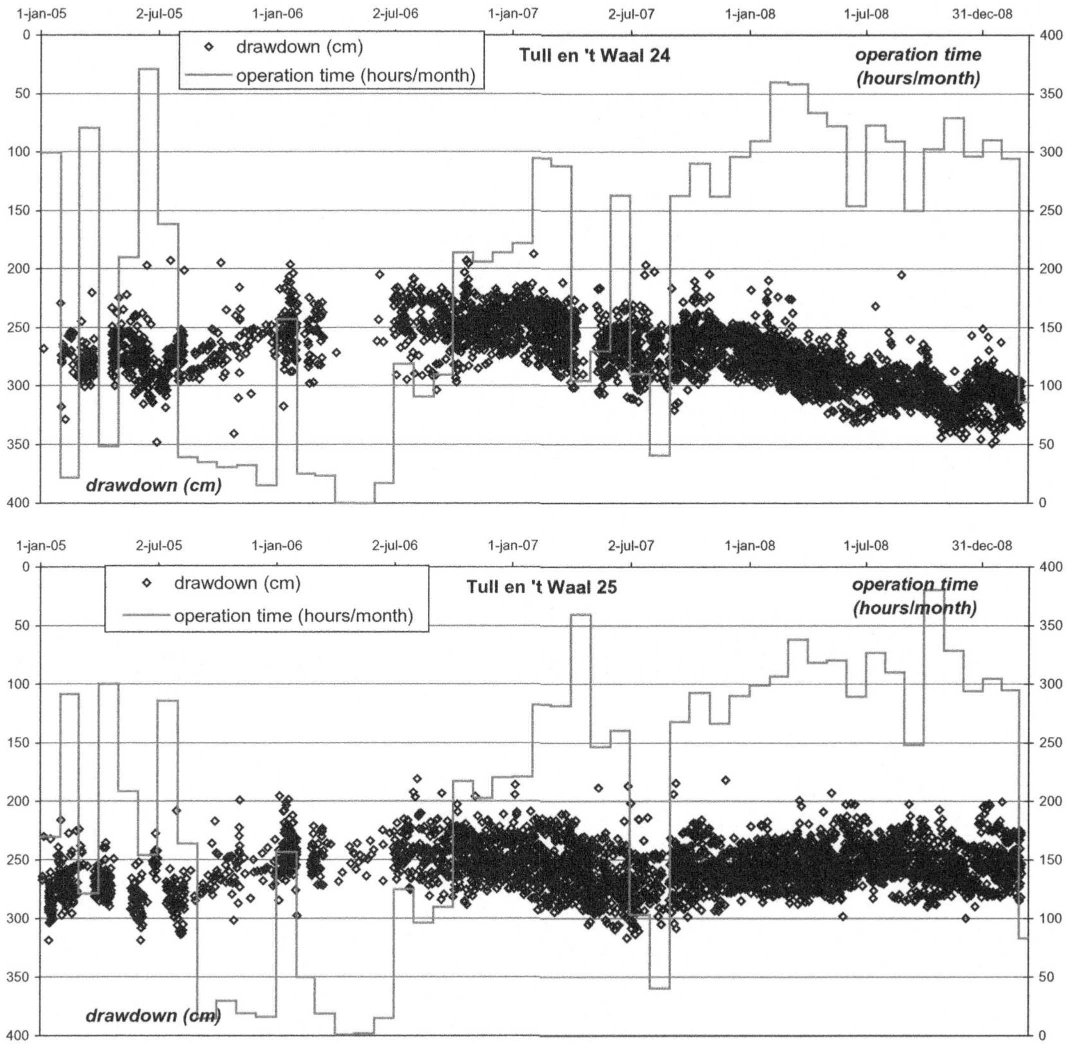

Figure 6.4 Drawdown and operating hours as function of calendar date for wells 24 (upper part) and 25 (lower part). Well 26 is not shown as the actual pressure exceeded the measurement range

Both other wells in this group improved until the autumn of 2006, after which drawdown in both wells increased. In well 24 this was around 75 cm over a period of 2 years, and in well 26 around 50 cm over a period of 2 years. This behavior is more in line with expectations: if drawdown increased with around 280 operating hours/month, it should also increase with around 300 hours/month.

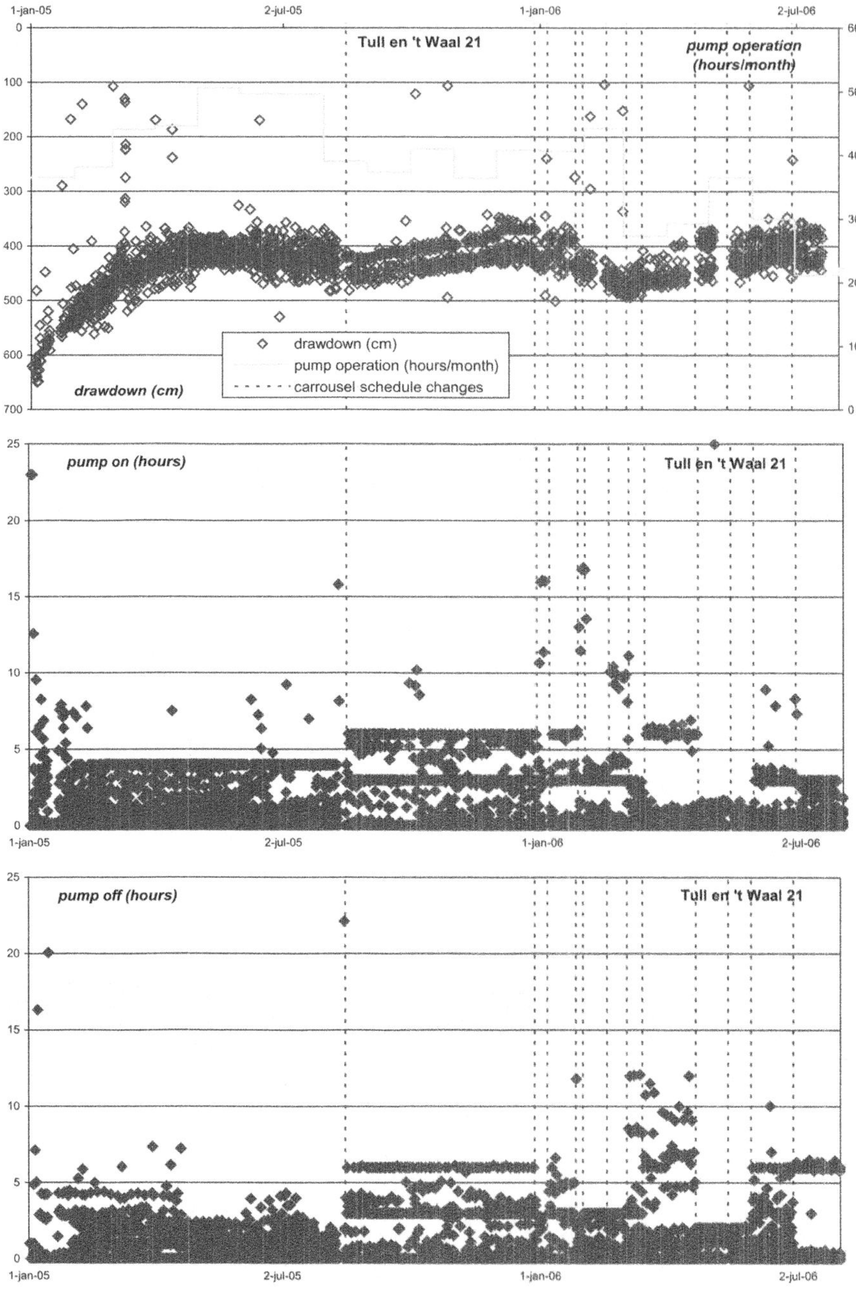

Figure 6.5 Drawdown, number of operating hours per month, changes in the carrousel schedule, and lengths of individual abstraction and rest periods as a function of calendar date for well 21

Additional influences on process operation

The preceding section demonstrated that intermittent abstraction controlled well bore clogging. However, it was also noted that the reaction to intermittent abstraction was not always unequivocal: wells 25 and 33 showed some clogging during periods with fewer production hours/month compared with periods when drawdown remained constant. Apparently, there are additional influences on drawdown. This is shown in Figure 6.5.

Drawdown in well 21 decreased by about 2m during the first 4 months of 2005. In May 2005, abstraction was increased to about 500 h/month, resulting in constant drawdown. This increased abstraction was accompanied by shorter individual rest periods. From 17 August 2005 onwards, drawdown decreased, but this decrease suddenly reversed to an increase on 28 December 2005, while production hours remained constant. Apparently, production hours/month is not a sufficiently distinguishing parameter, because the same number can represent very different distributions. Abstraction during 240 hours per month of 720 hours may be achieved by 24 hours abstraction followed by 48 hours rest, 12 hours abstraction followed by 24 hours rest, 8 hours abstraction followed by 16 hours rest, and so on.

Referring to these individual abstraction and rest periods, it appears that this reversal coincided with longer abstraction and shorter rest periods. The negative trend in drawdown was reversed again on 3 March 2006, coinciding with a decrease in pump operation. This decrease was characterized by longer individual rest periods and shorter individual abstraction periods. Apparently, in order to prevent well bore clogging, a given maximum time length of abstraction may not be exceeded, and/or a given minimum time length of rest is necessary.

Notable in Figure 6.4 are the very short abstraction and rest periods, sometimes less than 1 minute. These short switching cycles are due to a flaw in the current automatic operation system. As soon as the extension/renovation of the drinking water treatment plant is completed, the system will be reprogrammed, and this flaw will be eliminated.

6.4 ADDITIONAL EFFECTS BETWEEN WELLS

The load of particles accumulated on the well bore during abstraction may be represented as (Chapter 5):

$$B_{area,wb} = \frac{Q}{\pi dL}(c_{aquifer} - c_{wellhead})\Delta t \tag{6.1}$$

with:

$B_{area,owb}$: specific particle accumulation on well bore (n/m^2),
Q: abstraction rate of the (submersible) pump (m^3/h),
d: diameter of the well bore (m),
L: length of the well screen (m),
$c_{aquifer}$: background particle concentration in groundwater in the aquifer (n/m^3),
$c_{wellhead}$: particle concentration in abstracted groundwater (n/m^3),
Δt: length of abstraction period (h).

Equation (6.1) represents a measure for the susceptibility of a well for well bore clogging, where the difference between particle concentration in groundwater in the aquifer and in abstracted groundwater represents the "concentration" of particles retained on the well bore. From Equation (6.1) it can be derived that the larger the abstraction rate, the smaller the diameter of the well bore, the shorter the well screen, the larger the concentration of particles on the well bore, or the better the retaining properties of the well bore, and the longer the abstraction period, the greater the susceptibility for well bore clogging.

6.5 DISCUSSION AND CONCLUSIONS

Regularly switching on and off the pump prevents well bore clogging. This is clearly demonstrated by the examples presented in the preceding section, where drawdown increased, remained constant or decreased as a result of changes in well operation. This method of alternating operation and rest applies only to wells abstracting anoxic (sulfate reducing) groundwater. Prevention of screen slot (chemical) clogging requires exactly the opposite method of operation.

Wells react immediately to a large change in the method of operation (favorable or unfavorable) by a change (decrease or increase) in drawdown. Small changes may result in a delayed reaction. Apparently, there is an equilibrium between the method of operation and drawdown.

Not all wells react identically to an imposed system of alternating operation; the same schedule can increase drawdown in some wells and not in others. This is not surprising: abstracted groundwater from different wells shows different particle concentrations, and even if the particle concentrations were equal, the concentration might vary, not only over the length of the well screen, but also in a radial direction. Moreover, aside from particle concentration, the retaining properties for particles of the well bore vary between wells due to differences in drilling history and well development. Besides these factors, there may also be differences in well construction, capacity of the pump, and so on.

Prevention of well bore clogging is largely determined by the regular alternation of operation and rest. However, with comparable use, the same well sometimes clogs and sometimes does not. Apparently, there are some minor effects, where a maximum in the abstraction period may not be exceeded, and a minimum for the rest period is necessary.

The most unfavorable method of operation is continuous abstraction. Hoogendoorn et al. (2005) describe a non-clogging well field, where drawdown was not even monitored. Some time after a change from discontinuous to continuous abstraction necessitated by changing the water treatment method from aeration/filtration to membrane filtration, all wells were clogged and had to be replaced. In civil engineering projects as well, there may be continuous abstraction, resulting in conditions very susceptible to well bore clogging.

During pump tests executed on new wells and after rehabilitation of clogged wells, there may be a long, continuous abstraction period. The preceding experiences indicate that these long, continuous abstraction periods can result in well clogging. For that reason, it is advisable to not abstract for more than 30 minutes per day after rehabilitation of mechanically clogged wells. This schedule will also suffice to restore the bacteriological reliability of the abstracted groundwater.

Alternating abstraction and rest can be conveniently achieved with a carrousel operation system. The carrousels currently have very different operating hours; it may therefore be advisable to also rotate the carrousels. In this way wells in heavily used carrousels can recuperate when rotated to a lightly used carrousel. Of course, it is also possible to use all carrousels more evenly, or to redistribute the composing wells between the carrousels: all wells that are not very susceptible to clogging can be placed together in a heavily used carrousel and all wells susceptible to clogging in another, lightly used carrousel.

As a result of a flaw in the automatic operation system, wells were switched on and off very frequently, sometimes within one minute. This frequent switching is thought to exacerbate well bore clogging by disturbing the rest and abstraction periods.

Chapter 7

Prevention of two types of well clogging requires two types of well operationr*

C.G.E.M. van Beek[1], R.J.M. Breedveld[2] and P.J. Stuyfzand[1,3]

[1]*KWR Watercycle Research Institute, P.O. Box 1072, 3430 BB Nieuwegein, the Netherlands, +31 30 60 69 511, fax +31 30 60 61 165, Kees.van.Beek@Kiwa.nl*

[2]*Water utility Vitens*

[3]*Vrije Universiteit Amsterdam*

ABSTRACT

Wells abstracting groundwater are often susceptible to screen slot (chemical) clogging and wellbore (mechanical) clogging. Well clogging causes a decrease of a well's specific capacity over time and may threaten its reliability. Both types of clogging can be easily diagnosed during operation with the help of entrance resistance – the difference in head between the abstraction well and the observation well in its gravel pack. An increase in entrance resistance over time indicates screen slot clogging. The increasing resistance is caused by accumulation of chemical precipitates in the screen slots and adjacent gravel pack. Well bore clogging, however, does not show an increase in entrance resistance. Both types of clogging can be easily prevented, or at least mitigated, by two opposing methods of well operation: continuous abstraction for wells susceptible to screen slot clogging, and discontinuous abstraction (alternating operation with idle periods) for wells susceptible to well bore clogging.

7.1 INTRODUCTION

Groundwater is a reliable source for production of drinking water. However, wells used for abstracting groundwater are susceptible to clogging. Well clogging results in extra costs for monitoring, regeneration, and replacement of wells. In the Netherlands, these costs are estimated at an average of €5,000.=/year (≈US $ 6,500.=/year) for each clogged well.

*Published slightly modified as: C.G.E.M. van Beek, R.J.M. Breedveld and P.J. Stuyfzand (2009): Prevention of two types of well clogging. J. AWWA 101 (4) 125–134.

If clogging is not recognized in time, it may cause loss of production capacity, especially during periods of peak demand, when groundwater levels are deepest. Thus, well clogging may interfere with reliable delivery of drinking water.

This paper describes how to identify screen slot and wellbore clogging and how to prevent, or at least mitigate, both types of clogging. Regeneration methods, also distinct for each type of clogging problem (van Beek 1984) are not addressed in this paper.

7.2 IDENTIFICATION SCREEN SLOT AND WELL BORE CLOGGING

A decrease in a wel's specific capacity indicates either a screen slot (chemical) or well bore (mechanical) clogging problem. Specific capacity, defined as a well's abstraction rate divided by the drawdown, can decrease, even if the abstraction rate stays constant; that is, the well's drawdown increases.

Entrance resistance – the difference in head between the abstraction well and an observation well in the abstraction well's gravel pack – can help diagnose whether clogging is in the well's screen slots or on the well bore.

Screen slot clogging is characterized by an increasing entrance resistance over time. The continuous accumulation of chemical precipitates and/or biomass in the screen slots and adjacent gravel pack (Figure 7.1, right) causes the increased entrance resistance.

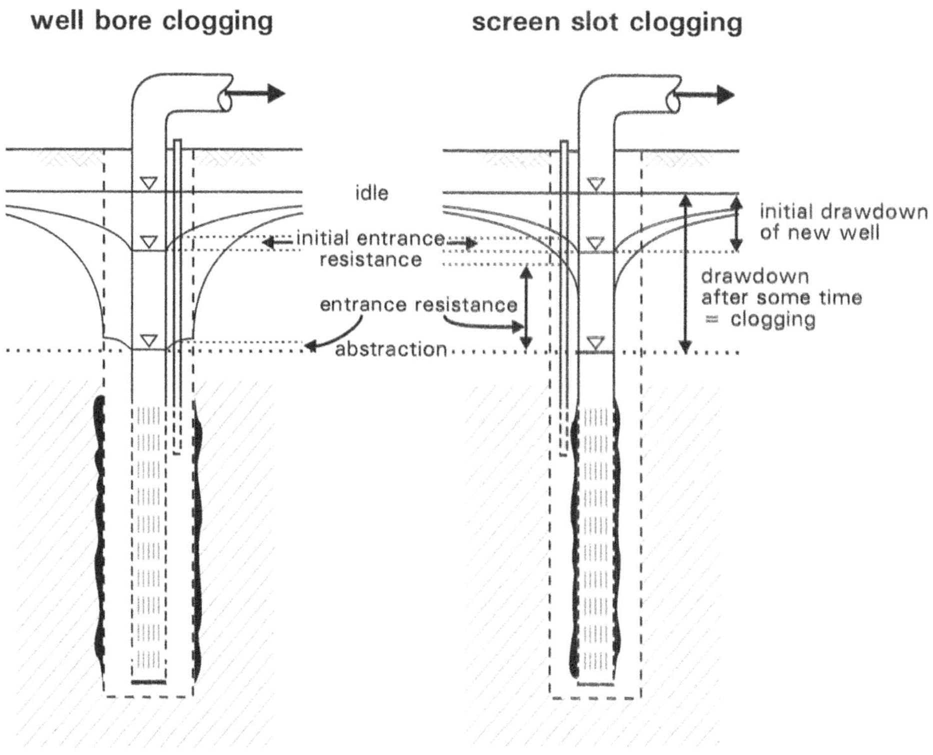

Figure 7.1 Differences in entrance resistance between well bore clogging (left) and screen slot clogging (right)

In contrast, well bore clogging is characterized by an entrance resistance that does not increase over time. Well bore clogging is caused by a continuous accumulation of particles on the well bore outside the gravel pack (Figure 7.1, left), and therefore does not affect entrance resistance.

Figure 7.2 shows field examples measuring drawdown in observation and abstraction wells affected by screen slot and well bore clogging. The well affected by screen slot clogging (Figure 7.2, left), shows that the difference in drawdown between the abstraction well and its observation well increases over time. However, in the well affected by well bore clogging (Figure 7.2, right), the difference in drawdown between the observation and abstraction well does not change.

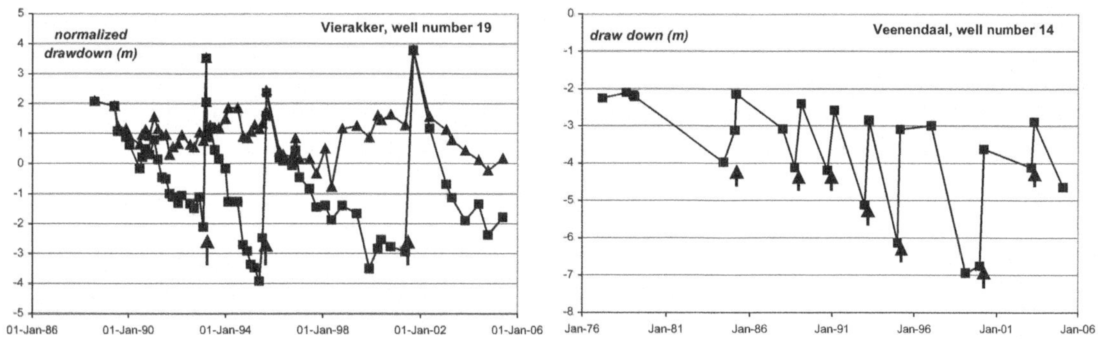

Figure 7.2 Field comparisons of drawdowns in observation wells and abstraction wells, for wells affected by screen slot (left) and well bore (right) clogging. Wells affected by screen slot clogging (left), show an increasing difference in drawdown over time between the observation well (upper line) and the abstraction well (lower line). In contrast, wells affected by well bore clogging, show no difference in drawdown between its abstraction well and observation well; the top line and bottom line coincide, showing no entrance resistance. Arrows represent regenerations

Screen slot clogging exhibits a number of additional characteristics. As chemical precipitates and biomass continue to develop after the abstracted groundwater has passed the screen slots, the precipitates will contribute to fouling of the submersible pump and water lines. Another characteristic is the chemical composition of the abstracted groundwater. Because the most common precipitate consists of iron hydroxides, abstracted groundwater often contains iron and oxygen. However, nitrate may be used as a surrogate parameter in testing because the presence of oxygen is sometimes hard to prove. A well's susceptibility to screen slot clogging by iron hydroxides precipitates can show up, even during drilling, by its combined abstraction from brown colored oxic sediments and gray colored anoxic sediments.

Well bore clogging exhibits rather different characteristics from screen slot clogging, including clean well screens, submersible pumps, and waterlines. Moreover, the chemical composition of the abstracted groundwater is characterized by the absence of oxygen (or nitrate, if used as a surrogate parameter), frequently showing sulfate reduction. Sometimes, however, the identification of well bore clogging may be confusing if the well is clogged so severely that it abstracts a mixture of water and air, resulting in the abundant formation of iron hydroxides, or the presence of iron hydroxides on the outside of the riser pipe.

Little exists in the literature about well bore clogging. However, screen slot clogging is well known and well documented in literature (Houben and Treskatis 2007; Roscoe Moss Co 1990; Driscoll 1986) and in many pictures (McLaughlan 2002). Well bore clogging is documented less because – as the authors found in Germany – it is not always considered a clogging phenomenon, but rather a delayed effect of improper well construction. However, careful review of the literature shows that well bore clogging is indeed a common problem, occurring throughout the world, for instance in Canada (Borch et al. 1993), Pakistan (Blackwell et al. 1995; Leake and Kamal 1990) and Germany (Grossmann 2000; Treskatis and Leda 1998; Berger et al. 1995; Witte 1981).

7.3 SCREEN SLOT (CHEMICAL) CLOGGING

7.3.1 Cause of screen slot clogging

For simplicity, this discussion focuses on screen slot clogging caused by mixing of oxygen- and iron-containing groundwater, whether or not microbiologically mediated. Well clogging can also be caused by manganese oxides, biomass of methane oxidizing bacteria, aluminum hydroxides and/or possibly calcite, and show similar characteristics.

Figure 7.3 shows abstraction of oxygen- and iron-containing groundwater from a phreatic aquifer in four abstraction phases, including the idle phases – phase 1, before the well is switched on, and phase 4, when the well is switched off.

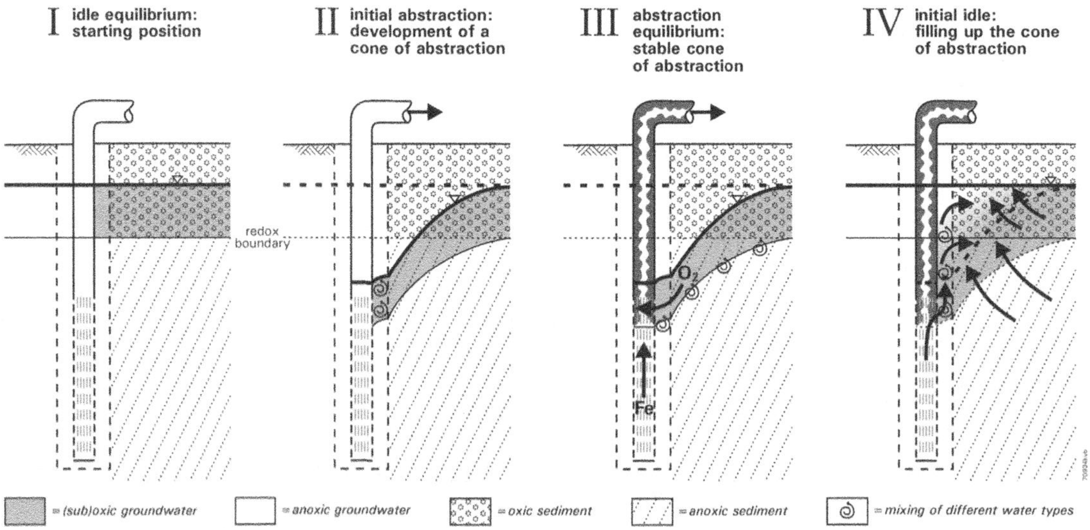

Figure 7.3 Four phases of groundwater abstraction from a phreatic aquifer

Phase 1. Idle equilibrium: Starting position. During idle equilibrium the well is not operating. In this example, the redox interface between oxygen-containing and iron-containing groundwater is some distance above the upper end of the well screen. This redox interface may be at any depth in the phreatic aquifer because the position of this interface is not essential.

Phase 2. Initial abstraction: Development of the cone of depression. After abstraction starts, a cone of depression develops in the water table. Consequently both the water table and the redox interface between oxygen and iron move down. This means that oxygen-containing groundwater enters anaerobic sediments, thereby oxidizing adsorbed iron to the iron precipitate that forms in phase 4, initial idle. In this example, even atmospheric oxygen enters anoxic sediments.

Phase 3. Abstraction equilibrium: Stable cone of depression. During abstraction, oxygen-containing water enters the well at the upper part of the screen, while iron-containing groundwater enters the well at the lower part. These two types of groundwater mix at the redox interface inside the well screen and further downstream in the submersible pump, riser pipe and raw water lines. Whether precipitates are formed in the screen slots in this phase depends on the interaction between the laminar flow entering the well through the slots and the turbulent flow upward inside the well screen, the residence time in the slots, and the reaction rate of iron hydroxide precipitation. The initial formation will probably be very slow, but as soon as precipitates are present, the reaction is accelerated because in addition to precipitation, adsorption of iron on iron hydroxide surfaces begins.

Phase 4. Initial idle: Filling up the cone of depression. After switching off the pump, the cone of depression starts to fill up. This in-filling may occur by upward and lateral flow of groundwater into the cone, and by upward flow of groundwater inside the well screen, leaving the well at the upper part of the screen and further upward through the coarse-grained, loosely packed gravel. If a clay seal has been installed in the bore hole above the well screen, this will reduce the upward flow via the gravel pack, depending on the length of this seal. During phase 4, mixed water inside the well screen has ample time for the formation of iron hydroxide precipitates. Moreover, iron flocs may swirl down from the submersible pump. This mixed water leaving the well will help fill the cone of depression back up, again providing ample time for the formation of iron hydroxide precipitates. The generated iron hydroxides will stick to the surface of the screen slots and the adjacent grains in the gravel pack. As soon as the surface is covered by iron hydroxide precipitates, the reaction rate will be enhanced because iron hydroxide deposits are generated not only by oxidation in solution, but also by coupled adsorption and oxidation.

In summary, the process of abstraction always consists of an inward flow through the slots of the screen with a very short residence time. These conditions are not favorable for the development of iron hydroxide precipitates. Rather, after the well is switched off there is ample time for iron hydroxides to form, and there is an initial outward flow gradient. During this outward flow iron hydroxide flocs may be retained by the slots of the screen as well as the adjacent gravel pack. This makes phase 4, the initial idle phase, decisive in initiating screen slot clogging. Once the iron hydroxide precipitates begin accumulating in the slots in phase 4, clogging will accelerate after the well is turned on again in phase 3 because these precipitates serve as nuclei for further precipitation.

7.3.2 Prevention of screen slot clogging

These findings led to the following recommendations for operating wells susceptible to screen slot clogging.

In order to prevent mixing of different water qualities, wells susceptible to screen slot clogging should be switched on and off as little as possible. Krems (1972) and Moser (1978) mention the relationship between well operation and well clogging, but are not as explicit as Wicklein and Steussloff (2006), who describe that in order to minimize chemical clogging the Berlin (Germany) Water Works operates its wells for lengthy abstraction periods.

Screen slot clogging may be minimized, not only by well operation, but also by well construction. This is discussed further in the following paragraphs.

Screen slot clogging may be mitigated by delaying the mixing of iron- and oxygen-containing water. For example, Kobus and Vlasblom (1975) present such an abstraction, as shown in Figure 7.4, by installing an inner pipe in the well screen that ends at about the interface of the oxygen-containing and iron-containing groundwater during abstraction. In this way, oxygen-containing groundwater is forced to flow downward, and iron-containing groundwater to flow upward, meeting each other at the entrance of the inner pipe (Figure 7.4, left), where the formation of iron hydroxide precipitates will start. A variation of this method

(Figure 7.4, right) is to close the bottom end of the inner pipe, forcing iron-containing groundwater to flow upward through the inner pipe and oxygen-containing water through the annulus between the inner pipe and the well casing.

These construction methods refer only to phase 3 of abstraction, abstraction equilibrium, and not to phase 2, initial abstraction, and phase 4, initial idle. Thus, these construction methods prevent formation of iron deposits in the well screen, and not in the screen slots and gravel pack.

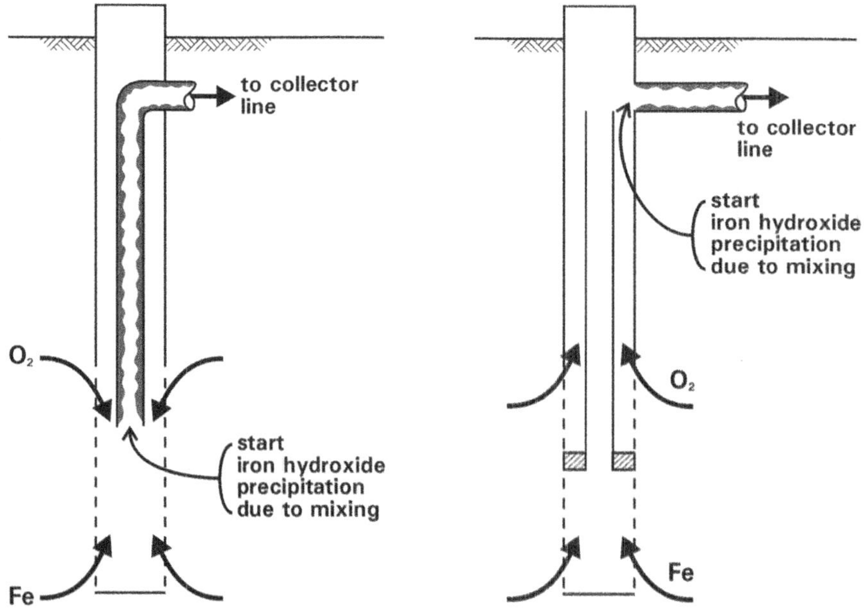

Figure 7.4 Delaying mixing of oxygen- and iron-containing groundwater to the bottom of the pipe (left) and to the top of the pipe (right). The example to the right shows a variation of the method of preventing iron hydroxide precipitates from clogging a well. The bottom end of the inner pipe is closed, forcing iron-containing groundwater to flow upward through the inner pipe and the oxygen-containing water through the annulus between the inner pipe and well casing

Another method of preventing iron deposits is to use shallow wells to abstract oxygen-containing groundwater, and deep wells to abstract iron-containing groundwater. Peters et al. (1992) adapted this approach for large numbers of small phreatic wells (1–10 m^3/h) abstracting groundwater by vacuum (Figure 7.5). This method of well construction achieves abstraction that is nearly separate and at the same time nearly continuous. Because screen slot clogging cannot be prevented completely, emphasis has been put on easy accessibility and easy regeneration. Another advantage of this type of well construction is that it is possible to measure the wellhead during abstraction, allowing the identification of the degree of clogging. This identification is not possible with traditional, closed well construction design.

Finally, another way to minimize screen slot clogging from iron hydroxide precipitates is application of subsoil iron removal (SIR), where iron is precipitated in the aquifer, so that the abstracted groundwater no longer contain iron (van Beek 1985).

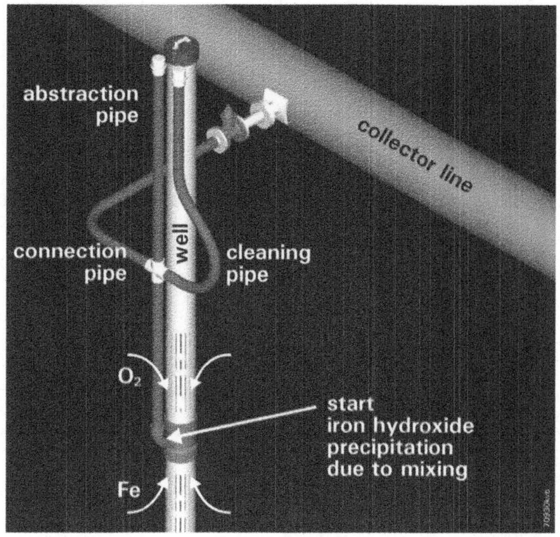

Figure 7.5 Abstraction of groundwater by vacuum to mitigate screen slot clogging (Source: P. Dammers, Dune Water Company, South Holland)

7.4 WELL BORE (MECHANICAL) CLOGGING

7.4.1 Cause of well bore clogging

Well bore clogging is caused by an accumulation of soil particles on the well bore. These particles are of geologic origin and are supplied to the well by the abstracted groundwater.

Using particle counters van Beek et al. (Chapter 3) demonstrated that abstracted groundwater contains 0.5 to >300 particles/mL with a diameter >2 μm. The larger-diameter particles, determined by the pore throats of the aquifer, were 20–50 μm. These particles consist of naturally occurring materials, like clay minerals, organic matter, quartz, and calcite (Timmer et al. 2003).

Figure 7.6 shows average particle concentrations calculated each minute during 27 h for well number 35 of well field Tull en 't Waal (water utility Vitens, Central Netherlands). Particles were counted online with the help of a particle counter connected to the sample valve on the well head.

Figure 7.6 shows large variations in particle concentration. Peaks in concentration are caused by switching on the submersible pump and by inadvertent causes, such as disturbance from switching on a nearby well. Switching the pump off also occasionally results in a peak in particle concentration. Apparently switching off causes some vibrations that mobilize particles present in the sample valve or sample tube. After the well is switched off, the raw water line remains under pressure and water from the water line starts flowing back through the particle counter. Because the raw water line is a big reservoir compared with the flow through the counter, the particle concentration remains constant.

As the pump is switched on, a double peak in concentration appears. The first peak coincides with switching on the submersible pump, and consists of particles that are mobilized in the sample valve by the corresponding vibrations. After a short time, a second but broader peak in particle concentration shows up. This time delay is equal to the standing volume of water in the well and gravel pack divided by the discharge of the submersible pump. Apparently, the second peak originates from a reservoir of particles on the well bore, accumulated during the preceding abstraction period. These particles are mobilized by several processes, for example by the collapse of particle bridges by absence of the flow of water toward the well or by the acceleration force generated by switching the pump on to set groundwater into motion in the aquifer.

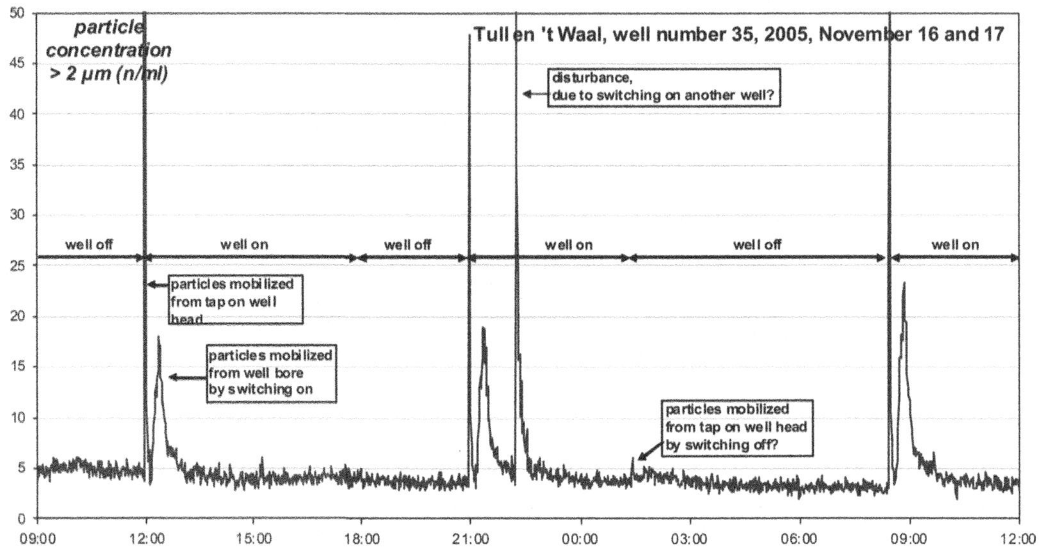

Figure 7.6 Particle concentration in abstracted groundwater during operating and idle periods from well number 35, well field Tull en't Waal (water utility Vitens, Central Netherlands). Particle concentrations were averaged for each minute during a 27-h period, a total of 1,620 counts on 16 and 17 November 2005. Concentrations were measured for particles >2 µm diameter

After this second particle-concentration peak has passed, the concentration of particles gradually decreases to a more or less constant value.

The well bore is usually considered a clean surface, but in practice there are remnants of drilling mud and fine sand and organic fragments from drilled upper layers resulting in constricted pore openings (van Beek 2002). Indeed, horizontal cores from well bores show accumulations of particles (Timmer et al. 2003) and drilling mud residue (de Zwart 2007).

Continuous measurements with pressure transducers in wells revealed that well operation can increase the occurrence of well bore clogging. Lengthy abstraction periods result in well clogging, while frequent switching on and off keeps the specific capacity of the well constant (Juhász-Holterman et al. 2004; Chapter 5). Other corroborating evidence for this interpretation includes examples like a non clogging well field becoming seriously clogged after operations changed from discontinuous to continuous abstraction (Hoogendoorn et al. 2005); and the disappearance of clogging after halving the production rate, or after installing smaller-capacity submersible pumps.

Both operational observations - a peak in particle concentration after the pump is switched on, and that continuous abstraction periods often result in clogging - can be understood with the help of particle balance data (Figure 7.7). During abstraction, not all particles are able to pass the well bore, and accumulate on the well bore and in the adjacent aquifer. After an idle period, when the well is switched on again, particles are mobilized. As long as, on average, the amount of particles removed by starting the submersible pump is equal to the amount accumulated on the well bore during the preceding abstraction period, no clogging will occur. If, on average, more particles stay on the wellbore than are removed, clogging is more likely to occur. This conceptual model explains why not all wells clog because in non clogging wells the equilibrium condition of accumulated and removed particles is being satisfied.

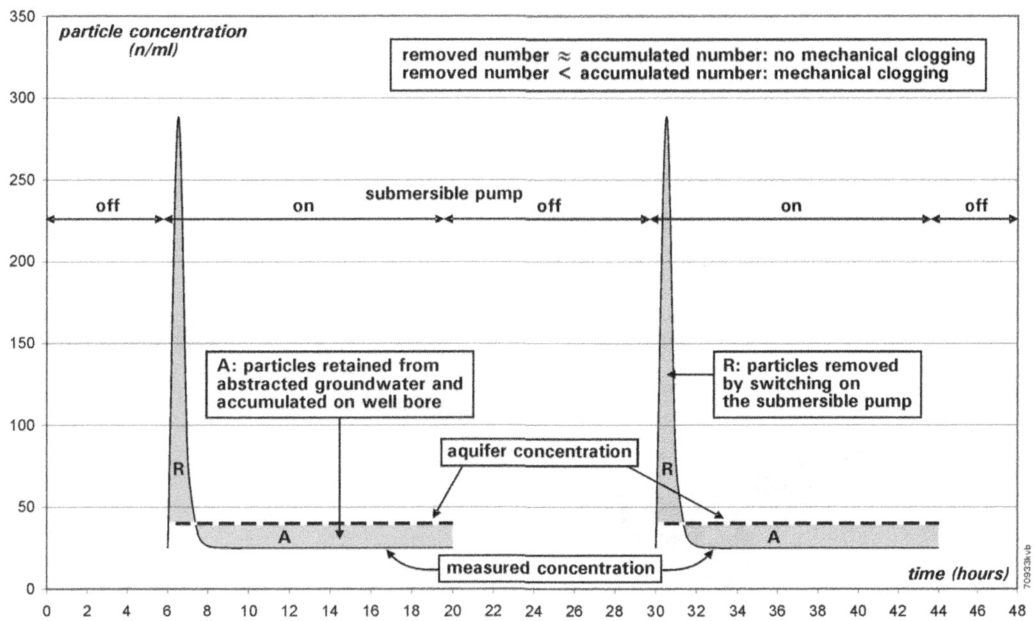

Figure 7.7 Schematic of well bore clogging. A: particles retained on the well bore, B: particles mobilized by switching the well on

7.4.2 Field testing

Well field Tull en 't Waal experiences serious well bore clogging. After limiting the length of the individual abstraction periods, the wells are not longer clogging and are even somewhat improving. Figure 7.8 shows results from testing the field's more than 30-year-old, often regenerated, well number 22. Well 22 was available for testing because it had to be abandoned for a planned extension of the treatment facilities. In order to gain direct information about mechanical clogging during well operation, the testing was started in such a way that the well would be likely to clog. During the testing period the water level in the well and well operation were monitored continuously. This was accompanied by some particle counting. Drawdown was calculated from these water level measurements.

Initially, well 22 ran continuously for up to 24 h. Later on, this maximum operation time was reduced manually to 12, 6, 3 and finally 1 h.

In August 2005, an automatic well operation program was introduced. From then on, the well ran up for 3 or 6 h. In April 2006 this pumping time was set to 6 h, and in May to 1 h. Later, the 3- or 6-h program was resumed.

Well 22′s operating condition and results of drawdown calculations are shown in Figure 7.8. As long as the abstraction rate remains constant, the drawdown is linearly proportional to the specific capacity of the well. The length of the idle period is also important, giving accumulated particles an opportunity to detach from the well bore.

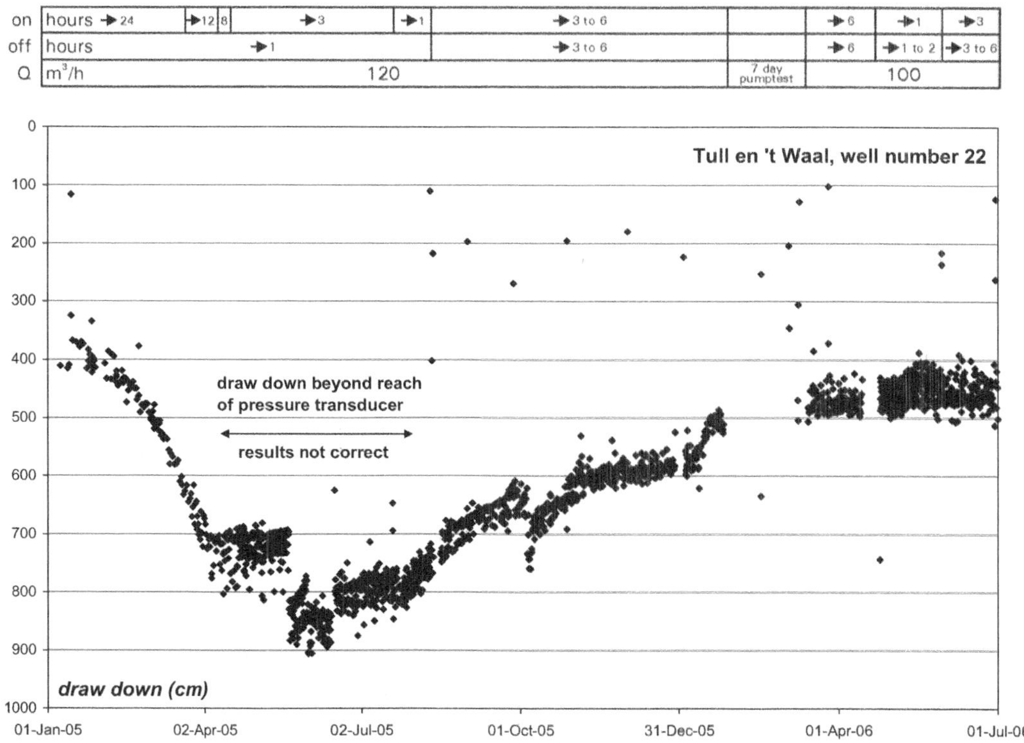

Figure 7.8 Well operation and drawdown in well 22 of well field Tull en 't Waal during the experimental period. (In well operation the abstraction hours are up to the programmed time lengths. It was not always possible to run the well the prescribed time length, as otherwise the drinking water reservoir would overflow.)

Figure 7.8 shows clearly that the drawdown initially increased sharply and after May 2005 gradually decreased, approaching its initial value. This initial sharp increase in drawdown coincided with long operation hours. After the length of operation time was shortened to 3 h, the increase in drawdown stopped and even began to recover. Unfortunately, this change is not directly visible. Extrapolation of the drawdown line from both sides shows that the turning point must have occurred soon after introduction of the 3-h operation scheme. During July-August 2005 and May 2006, a 1 h operation scheme was initiated. This scheme did not result in a greater improvement than the 3- to 6-h operation scheme.

During February 2006, a seven-day pump test was conducted on well 22. During this pump test, the well was run continuously with varying capacities up to 100 m^3/h. Figure 7.8 clearly shows that this continuous abstraction did not result in a deterioration of the well. Apparently, a "healthy" well may be overcharged for some time without any harm.

This equilibrium condition has been corroborated by results of particle counts; that is, the number of particles mobilized when the submersible pump was switched on appeared comparable to the number of particles accumulated during abstraction (Chapter 5).

7.4.3 Prevention of well bore clogging

These findings led to the following recommendations for managing wells susceptible to well bore clogging. In order to remove accumulated particles from wells susceptible to well bore clogging, the well should be switched on and off regularly. The length of the abstraction and idle periods is important in managing wells susceptible to well bore clogging and is determined by the characteristics of the site and of the well itself. Important considerations are the well's particle load, pump, and the filter capacity of the well bore. Remnants of drilling mud and trapped fine sand and organic fragments left over from the original drilling and well construction process, affect the well bore's filter capacity (Chapter 5).

The particle load depends on the particle concentration, the surface area of the well bore, and the abstraction rate; the abstraction rate also affects the particle concentration. Therefore, wells with a larger-diameter well bore and wells installed with lower-capacity pump tend to be less susceptible to well bore clogging.

The retaining properties of the well bore are determined by the diameter of the pore constrictions. The larger the pore spaces, the more easily mobilized particles may pass. Reducing a well's particle retention can be achieved during the original well drilling of the well; that is, by completely removing drilling mud, and by installing as coarse a gravel pack as possible.

7.5 APPLICATION

The testing mentioned in this paper was accomplished with the help of well fields used by the Vitens water utility. Table 7.1 presents a summary of relevant average chemical characteristics of the abstracted groundwater from 12 of the Vitens fields.

Table 7.1 Characteristics of well fields exploited by the Vitens water utility in the Central Netherlands

Well field	SO_4 mg/l	NO_3 mgN/l	Fe mg/l	Mn mg/l	NH_4 mg/l	Redox state	Observed well clogging
Beerschoten	17	<0.05	2.25	0.22	0.19	anoxic	no clogging
Berg	25	3.30	<0.01	<0.01	<0.02	oxic	no clogging
Bilthoven	24	2.30	0.35	0.17	0.05	mix	screen slot clogging
Bunnik	12	<0.05	2.65	0.24	0.18	anoxic	well bore clogging
Cothen	<2.0	<0.05	2.95	0.18	0.15	anoxic	well bore clogging
Groenekan	22	<0.05	3.80	0.30	0.40	anoxic	no clogging
Laren	20	0.55	2.20	1.30	0.36	mix	screen slot clogging
Soestduinen	20	0.90	0.02	0.02	<0.02	mix	minimal screen slot clogging
Tull en 't Waal	<2.0	<0.05	0.80	0.13	0.17	anoxic	well bore clogging
Veenendaal	5	<0.05	0.45	0.06	0.31	anoxic	well bore clogging
Woudenberg	2	<0.05	0.54	0.05	0.28	anoxic	no clogging
Zeist	30	1.80	0.95	0.25	0.07	mix	screen slot clogging

From the data shown in Table 7.1 it appears that well fields abstracting only oxic groundwater, characterized by the presence of nitrate as a surrogate for the presence of oxygen, and by the absence of iron (i.e. Berg and Soestduinen well fields), do not clog. Well fields abstracting mixed redox groundwater, characterized by the presence of nitrate and of iron (i.e. Bilthoven, Laren and Zeist well fields) experience screen slot clogging.

Wells abstracting anoxic groundwater, characterized by the absence of nitrate (oxygen), are susceptible to well bore clogging. However, not all wells clog; it might be that these wells by luck are operated in such a way that no clogging occurs. Moreover, these non-clogging wells abstract groundwater from a coarse-grained, leached aquifer, containing no large pool of particles. This low particle concentration makes the wells less susceptible to mechanical clogging, which allows more freedom in well operation.

Surprisingly, wells showing screen slot clogging can easily be regenerated to their original capacity. This means that the anoxic part of the well screen abstracting iron-containing groundwater, shows no wellbore clogging. The reason for this phenomenon is not known. It could be that particles are mobilized more easily under sulfate-reducing conditions than under iron-reducing conditions, or that the drilling mud and drilling additives are reacting differently under sulfate- and iron-reducing conditions. Another possible explanation is that these wells are operated so that clogging is minimized. Usually, these wells are somewhat smaller than wells abstracting anoxic groundwater, and are probably switched on and off more frequently.

7.6 CONCLUSIONS

Well clogging is characterized by a decrease of the well's the specific capacity over time. Two types of well clogging can occur: screen slot (chemical) clogging, characterized by an increasing entrance resistance over time, and well bore (mechanical) clogging, characterized by the absence of an entrance resistance. Other indicators include the occurrence of, or lack of, fouling in the pump intake and water discharge lines, and natural water quality.

Both types of clogging can be prevented, or at least mitigated, by two opposing methods of well operation: continuous abstraction for wells susceptible to screen slot clogging, and alternating abstraction and idle periods for wells susceptible to well bore clogging.

Screen slot clogging also can be prevented by adaptations in well construction; for example, by separate abstraction of oxygen and of iron by separate wells. Clogging may be mitigated by the installation of an extension pipe under the pump, delaying mixing of both water types to the end of this extension pipe. This adaptation only prevents clogging during abstraction, not after abstraction. Moreover, it moves the beginning of the precipitation of iron hydroxides to the end of this extension pipe, where both water types will start to mix. Screen slot clogging may also be minimized by the application of external methods, such as SIR technology.

In preventing well bore clogging by regularly alternating abstraction and idle periods, it appears that not only the length of the abstraction period is important, but also the length of the idle period. This alternation of abstraction and rest may be conveniently implemented with the help of an automated well operation scheme. The length of the alternating periods is determined by the particle load, and thus by the capacity of the submersible pump, and the filter capacity of the well bore. The drilling mud remnants and trapped fine sand and organic fragments left over from the original drilling and well construction process, affect the well bore's filter capacity.

The particle accumulation model justifies extending well development efforts beyond that of merely visual inspection of sand-free water. It also makes a case for less conservative (coarsely packed) gravel pack designs to help minimize trapping of particles.

Chapter 8

Well incrustation and well design criteria

C.G.E.M. van Beek[a,*] and P.J. Stuyfzand[a,b,1]

[a]*KWR Watercycle Research Institute, Groningenhaven 7, 3433 PE Nieuwegein, the Netherlands*

[b]*VU-University, De Boelelaan 1105, 1081 HV Amsterdam, the Netherlands*

[1]*Tel.: +31 30 606 9511; fax: +31 30 606 1165 and tel.: +31 20 598 9898; fax: +31 20 598 9899*

[*]*Corresponding author. Tel.: +31 30 606 9511, fax: +31 30 606 1165.* E-mail adresses*: Kees.van.Beek@kiwa.nl (C.G.E.M. van Beek), Pieter.Stuyfzand@kiwa.nl and Pieter.Stuyfzand@falw.vu.nl (P.J. Stuyfzand)*

ABSTRACT

Well clogging due to precipitation of iron hydroxides is governed by the simultaneous abstraction of oxygen-containing and of iron-containing ground water and by well operation; it is not or only marginally influenced by turbulent flow. Precipitation of calcium carbonate caused by degassing of carbon dioxide is governed by the total gas concentration and the depth below the water table, and is not or only marginally influenced by turbulent flow. In addition, current well design and well operation result in laminar groundwater flow, or at the most incipient turbulence. However, contrary to expectations, well clogging due to accumulation of iron hydroxides is still a frequent problem. Maintaining laminar groundwater flow is therefore not a relevant well design parameter, indicating that higher abstraction rates may be applicable. Instead, each situation should be analyzed individually to determine which clogging process might be relevant, if any, and attune well design to these circumstances. Application of this reasoning leads to site-specific well design and well operation rules.

8.1 INTRODUCTION

Basic principles for the design of groundwater abstraction wells are included in many handbooks (Houben and Treskatis 2007; DVGW 2005; Lehr et al. 1988; Driscoll 1986; Roscoe Moss 1990; Smith 1992). These principles include:

a. Minimal drawdown during abstraction. This principle is obvious and needs no further discussion.
b. Minimal suffosion (=erosion and transport) of particles. Suffosion should not be confused with sand supply by a well. Sand supply is usually ascribed to an incorrect grain diameter of the gravel pack, to an inappropriately installed gravel pack, due to bridging or a poorly centered well screen, or to a settling gravel pack, especially where a clay seal has been applied just above the top of the well screen. Problems with a settling gravel pack can be avoided by continuously supplying gravel using a feed pipe (Roscoe Moss 1990; Howsam et al. 1995) or starting the top of the screen some distance below the aquitard (Tholen 2006).
c. Minimal colmation (or colmatation), i.e. accumulation of particles on (outer colmation) or in (inner colmation) a filter (gravel pack) leading to clogging.
d. Minimal degassing, in particular of carbon dioxide.
e. Minimal mixing of incompatible waters. Minimal degassing and minimal mixing of incompatible waters are often discussed together as minimizing incrustation (Houben and Treskatis 2007; Bieske et al. 1998; Driscoll 1986).
f. Minimal corrosion. In case of PVC wells, corrosion is irrelevant, and will not be considered further.

All these principles are related to groundwater flow velocity, either as result of turbulent flow, like degassing and mixing, or through transport of fine material, like erosion, colmatation and suffosion. According to Bieske et al. (1998) and Ohio (2000), laminar flow should prevent incrustation.

Here, first the conditions for well clogging due to accumulation of iron hydroxide precipitates will be presented. Next, the conditions for degassing of carbon dioxide from groundwater in relation to calcite precipitation will be discussed. After this the occurrence of laminar and turbulent flow in unconsolidated sedimentary aquifers in and around groundwater abstraction wells, i.e. in the aquifer, gravel pack, screen slots and well casing will be demonstrated. These parts of the paper provide the information to evaluate the relevance of turbulent groundwater flow to chemical well clogging. Finally, the current rules for well design will be discussed. These rules appear too conservative; depending on the vulnerability to clogging, higher than currently applied abstraction rates should be possible.

8.2 CLOGGING BY ACCUMULATION OF IRON HYDROXIDE PRECIPITATES

8.2.1 Characteristics

Screen slot (or chemical) clogging is characterized by accumulation of precipitates in the screen slots (Houben and Treskatis 2007; McLaughlan 2002; Driscoll 1986). However, these precipitation processes do not come to an end in the slots but continue further downstream, in particular on the screen of the submersible pump and inside the riser pipe and collector lines. These precipitates consist of iron hydroxides and manganese oxides whether or not accompanied by slimes (biomass). Screen slot clogging may also occur by accumulation of other precipitates like aluminum hydroxides (Baudisch 1989) and calcite but these will not be considered here.

As precipitation of iron hydroxides is most prevalent, discussion is limited to this process. The precipitation of iron hydroxides requires oxygen, which cannot be provided by diffusion over the water-gas interface inside the well: the area of the water-gas interface inside the well, i.e. the annulus between the well casing and the riser pipe, is too small and diffusion over the water/gas interface too slow (Stumm and Morgan 1996) to be relevant. However, the presence of atmospheric oxygen inside the well may result into iron precipitates outside on the riser pipe over the range of the fluctuating water table. Sometimes this occurrence is misinterpreted as the presence of screen slot clogging.

Oxygen, necessary for the oxidation process, must therefore be provided by the abstracted water. Infiltrating precipitation water is saturated with respect to oxygen. As soon as this water enters the soil, oxygen may be used for root respiration and mineralization of organic matter, resulting in a decreasing concentration of oxygen and an increasing concentration of carbon dioxide. By diffusion and convection (by diurnal differences in temperature at the soil surface) the soil gas composition tries to restore the original atmospheric composition. Depending on local conditions, in particular groundwater depth, the concentration of oxygen in precipitation water recharging groundwater may be lower than saturated with respect to atmospheric conditions. Depending on the reduction capacity of the aquifer matrix, dissolved oxygen will penetrate more or less deeply. Actually, a redox reaction sequence is started (Stumm and Morgan 1996; Appelo and Postma 2006): consumption of oxygen, denitrification (if nitrate is present), dissolution of manganese and iron, sulfate reduction and methane production.

Figure 8.1 shows that part of the redox sequence as function of depth, that is relevant for iron hydroxide precipitation: the successive disappearance of oxygen and nitrate and the appearance of iron. This figure is characterized by a ca 5 m deep inert depth interval not containing oxygen but nitrate and by a sharp nitrate-iron redox interface. If a well is sunk in this aquifer, a mixture of oxygen containing water from the upper part and iron containing water from the lower part of the aquifer will be abstracted, and clogging will develop inevitably.

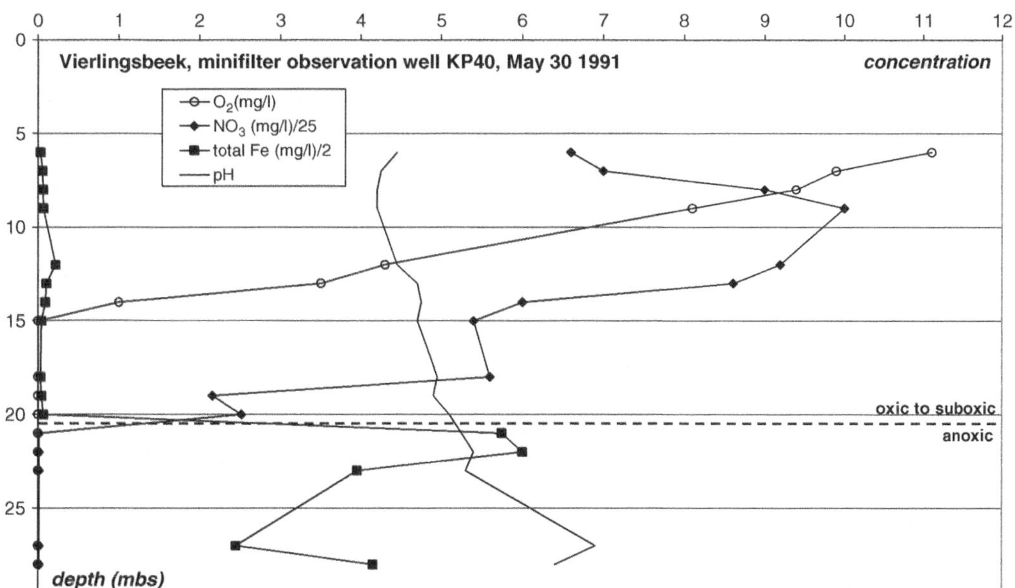

Figure 8.1 Chemical composition of groundwater over depth, measured with the help of a multilevel minifilter monitor well near Vierlingsbeek (water utility Brabant Water). Oxygen concentration measured 17 February 2009 (Courtesy Harry Boukes)

From the above reasoning several characteristics of screen slot (or chemical) clogging by iron hydroxides may be derived. A chemical characteristic says that the abstracted water must contain oxygen and iron, however both concentrations in the abstracted, mixed, water are lower than in their stable part of the aquifer, see also Stuetz and McLaughlan (2004). Complete chemical analyses of ground water usually contain the concentrations of nitrate and iron, and not of oxygen. Often nitrate is used as a surrogate parameter for oxygen: if the abstracted groundwater contains nitrate, oxygen is usually present too; if the abstracted water contains no nitrate, oxygen may be present, as nitrate may be considered a pollutant.

Moreover a landscape characteristic may be derived: in geologically young sediments, the presence of oxygen is limited to phreatic aquifers. Consequently the occurrence of screen slot clogging is limited to phreatic aquifers. In old sedimentary environments, oxygen may be present in deeper aquifers, and so the occurrence of screen slot clogging.

The iron hydroxide precipitates accumulate in the screen slots, and hamper in this way the flow of water through these slots. This blockage results in an increasing entrance resistance, which provides a hydraulic characteristic for screen slot clogging.

Aerobic (oxic) sediments contain ground water with dissolved oxygen and are (reddish) brown colored; anaerobic (anoxic) sediments contain ground water with dissolved iron(II) and are (whitish) gray colored. The transition between both environments may be mottled (gley zone), pointing towards alternating aerobic and anaerobic conditions. This transition zone contains iron oxidizing bacteria (Chapelle 1992). From this observation a geological characteristic may be derived: wells abstracting ground water simultaneously from brown and from gray colored sediments will clog. So already during drilling, geology warns for the occurrence of screen slot clogging.

The role of nitrate as an oxidator of ferrous iron is not completely clear. Chemical oxidation of ferrous iron by nitrate asks for rather extreme conditions (Buresh and Moraghan 1976). Recently Straub et al. (1996) demonstrated the presence of microbiologically mediated oxidation of ferrous iron by nitrate (Straub et al. 2001). In the field, oxidation of ferrous iron in iron sulfides by nitrate had been observed frequently (Kölle et al. 1990; van Beek et al. 1989). Moreover, Kölle and Schneider (1992) showed that in Subsoil Iron Removal (SIR) nitrate is able to oxidize adsorbed iron.

8.2.2 Screen slot (or chemical) clogging

If a well abstracts a mixture of oxygen-containing groundwater from the upper part of the aquifer and iron-containing groundwater from the lower part, iron hydroxides will start to precipitate. Surprisingly, these precipitates are not only present in the screen slots and further downstream, but also somewhat upstream in the gravel pack (Houben 2006; Houben and Treskatis 2007), where mixing between both water qualities during abstraction is negligible.

In order to illustrate the iron hydroxide precipitation processes in and around well screens, leading to screen slot clogging in phreatic aquifers, 4 phases may be distinguished: a. Idle equilibrium, b. Initial abstraction, c. Abstraction equilibrium, and d. Initial idle. The differentiation into four phases is shown in Figure 8.2 and further explained below.

Phase I: Idle equilibrium.

In this phase there is no abstraction and, consequently, no drawdown. Figure 8.2 shows this phase for a phreatic aquifer. In this example, the redox interface between groundwater containing oxygen and groundwater containing iron is present at some distance above the upper end of the well screen. This redox interface may be present at any depth in the phreatic aquifer, see also Figure 8.1, as the position of this interface is not essential.

Phase II: Development of the cone of depression.

As soon as abstraction starts, a cone of depression will develop around the well in the water table. In Phase II in Figure 8.2, both the water table and the oxygen-iron redox interface in the groundwater move down. This means that oxygen-containing groundwater enters anoxic sediments. In our example, even atmospheric oxygen enters anoxic sediments.

During this passage of oxic groundwater through anoxic sediments, oxygen oxidizes all kinds of reduced compounds present in these anoxic sediments, such as adsorbed iron(II). As a result, adsorbed iron(II) changes into fresh (active) iron hydroxide.

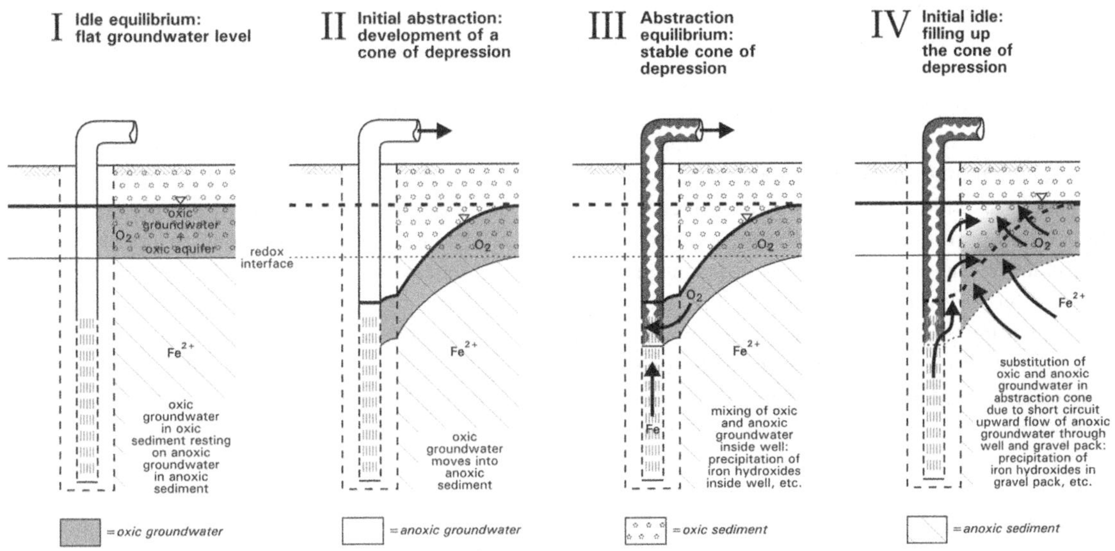

Figure 8.2 Phases identified during abstraction of oxygen containing and iron containing groundwater from a phreatic aquifer

Phase III: Abstraction equilibrium.

In this phase, see Figure 8.2, the well has developed a more or less stable cone of depression, where oxygen-containing water enters the well above the redox interface in the upper part of the screen, and iron-containing water below this interface in the lower part. As the upward flow in the well is turbulent, these two types of groundwater mix intimately from the redox interface on in the well and further downstream.

In a brand-new clean well, the only process active in iron hydroxide precipitation is homogeneous oxidation. Under neutral conditions this process is rather slow (Chapter 2). As the residence time in the slots is short, homogeneous iron hydroxide precipitation in the slots will be negligible. Moreover, most of the iron hydroxide flocs, generated in this way, will be carried away and may precipitate on the screen of the submersible pump and further downstream. However, as soon as precipitates are present in the slots, the reaction rate is accelerated, as besides homogeneous oxidation also heterogeneous oxidation becomes active (Chapter 2).

Iron oxidizing bacteria are present everywhere, and will start to thrive as soon as conditions are favorable. These conditions are excellent above the redox interface inside the well, where the water contains oxygen as well as iron and where (rough) screen slots may serve as excellent settling locations, accelerating the oxidation reaction rate (Ralph and Stevenson 1995).

The contribution of shallow oxygen-containing groundwater to the total abstraction depends on the position of the redox interface relative to the well screen and the eventual presence of a clay seal above the gravel pack or confining layers in the aquifer. The absence of a clay seal above the well screen results in a short-circuit flow of groundwater through this pack from the upper part of the aquifer to the upper part of the well screen.

Phase IV: Filling in of the cone of depression.

After switching off the pump, the water level inside the well (drawdown) and outside the well (cone of depression) will start to move back to their idle position. Because the gravel pack contains larger grains and the bore hole has a less compact structure than the surrounding aquifer, the vertical permeability of this pack is larger than that of the aquifer, and in order to fill in the cone of depression an upward flow from the well through the gravel pack may arise. Water in the well initially consists of a mixture of oxygen-containing and iron-containing groundwater. In this mixture there is ample time for oxidation and iron hydroxide precipitation. Moreover, iron hydroxide flocs may twirl down from the screen of the submersible pump. These flocs will be carried along by the water filling in the cone of depression and precipitate on the screen slots and adjacent gravel pack (Houben and Treskatis 2007). In particular if no clay seal has been applied on top of the gravel pack, the volume of water leaving the upper part of the well screen into the gravel pack (Phase IV in Figure 8.2) may be relevant. This water will be replenished by iron-containing groundwater from the deeper part of the well. In this way, the well screen acts as a short circuit conduit, and iron containing groundwater will pass an aerobic environment, resulting in adsorption of iron by iron hydroxide precipitates. This process of alternating preferential downward flow of oxygen containing and preferential upward flow of iron containing water results in alternating oxidation of adsorbed iron by dissolved oxygen and adsorption of iron(II) by fresh iron hydroxide precipitates and is identical to SIR in which oxygen-containing water is periodically infiltrated into wells delivering iron-containing groundwater (van Beek 1985, 1983, 1980; Appelo et al. 1999).

The cone of depression may also be refilled by upward and lateral flow of shallow groundwater (phase IV in Figure 8.2). The proportion of refilling between short circuit flow through the gravel pack and upward and lateral flow of shallow groundwater depends on the local flow conditions.

After the cone of depression has been refilled, phase I is reached again; and after resuming abstraction all phases are passed through again.

8.2.3 Precipitation of iron hydroxides in the aquifer

So far the precipitation of iron hydroxides in the screen slots and adjacent gravel pack, and downstream on the well screen, submersible pump etc. has been discussed. However, there are reports that recent iron hydroxide precipitates have been encountered further removed from the well in the aquifer (Tyrrel and Howsam 1997; Houben and Weihe 2010; Houben 2006). Several processes may explain this presence.

After switching off the submersible pump, the oxygen-iron redox interface in the aquifer will move back to its idle position. In case of a sharp oxygen-iron(II) interface, anoxic groundwater containing iron(II) passes sediments in which active iron hydroxides developed during the preceding abstraction period, now able to adsorb iron(II) from passing groundwater. In the next abstraction period, this adsorbed iron(II) is oxidized by oxic groundwater moving downward (see also Driscoll 1986). Consequently, as a result of alternating abstraction, the oxygen-iron(II) interface will move up and down in the aquifer. This fluctuation of the oxygen-iron(II) interface goes along with alternating oxic and anoxic conditions, resulting in heterogeneous iron hydroxide precipitation. This process has been schematized in Figure 8.3.

The presence of fresh iron hydroxide precipitates at some distance from an operating well may also be explained by the presence of differences in potential over the height of the aquifer. If, during rest, there is present a difference in potential head between the upper and lower part of the aquifer, the well screen may act as a short circuit flow conduit. In case of a lower potential in the lower part of the aquifer, oxygen containing water may enter an anaerobic environment, and in the opposite case iron containing water an aerobic environment. During abstraction the original water quality will be abstracted. As a result, these alternating oxic and anoxic conditions may result in heterogeneous iron hydroxide precipitation, comparable to SIR. As expected (van Beek and Vaessen 1979), these heterogeneous oxidation precipitates are dense, consisting for the greater part of goethite (Mettler et al. 2001).

In the presence of a sharp oxygen-iron(II) redox interface, oxygen containing and iron containing water may mix during abstraction due to transversal dispersion. Sharp redox interfaces and large values for transversal dispersion (transversal dispersivity and groundwater flow velocity), result in intensive mixing. This mixing may result in homogeneous, heterogeneous and possibly biological iron hydroxide oxidation and precipitation within the aquifer.

8.2.4 Discussion

In conclusion, a prerequisite for well clogging by iron hydroxide precipitates is the simultaneous presence of iron and oxygen in the abstracted groundwater. If both components are present, clogging is primarily governed by well operation, and not by turbulent flow. Consequently, avoidance of turbulent flow is inadequate as a design criterion.

This explanation of the occurrence of screen slot clogging also presents a method for prevention: continuous abstraction will prevent clogging, and in particular the onset. However, continuous abstraction will not prevent deposition of iron hydroxide precipitates inside pumps and pipes. The favorable relation between prevention of clogging and continuous abstraction was already observed by Krems (1972), Moser (1979, 1978) and Tyrrel and Howsam (1994) and explicitly mentioned by Wicklein and Steußloff (2006). However, it must be realized that this rule applies to screen slot clogging, and not to well bore clogging (Chapter 7).

Prevention of screen slot clogging is also possible by disinfecting wells during one day/month with circa 1 g/l chlorine as sodium hypochlorite (Krems 1981, 1972; DVGW 1970) or with 2% hydrogen peroxide (Wicklein and Steußloff 2006). However, Moser (1978) was not always successful in preventing well clogging with chlorine. This experience shows that iron oxidizing bacteria often play a dominant role in screen slot clogging, but not always.

Sodium hypochlorite and hydrogen peroxide not only have a disinfecting effect, but also an oxidizing effect. In this way they may promote the oxidation of dissolved and adsorbed iron.

In case of a shallow redox interface, clogging will start at the top of the well screen, and will move slowly down, hampering in particular the flow of shallow groundwater to the well, as if the upper part of the well casing moves down. By consequence, the (relative) contribution of shallow groundwater to the total abstracted groundwater will decrease. Under conditions this may result in a decreasing oxygen and nitrate concentration or increasing iron concentration during increasing clogging. The shallow inflow of oxygen containing water is corroborated by downhole camera pictures (Tyrrel and Howsam 1994;

Hässelbarth und Lüdemann 1967) and results of strings of perspex slides, suspended in clogging wells (Kobus and Vlasblom 1975): slides in the upper part of the screen are covered by iron hydroxide deposits, while slides in the lower part remain clean.

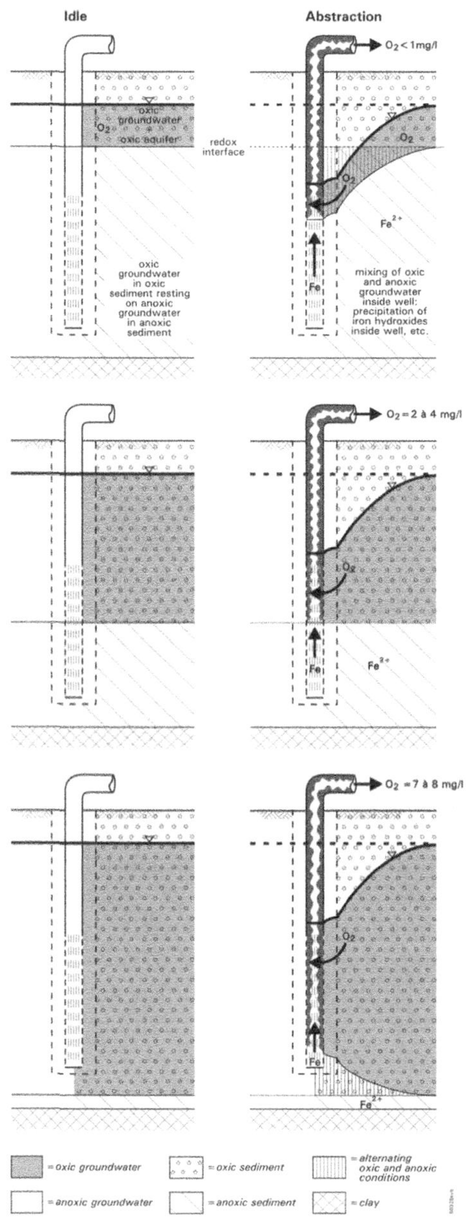

Figure 8.3 Development of iron hydroxide precipitates in the aquifer by heterogeneous oxidation as function of the position of a sharp oxygen-iron(II) redox interface. Iron hydroxides will only precipitate in the upper and lower representation

If the oxygen-iron redox interface is present halfway the aquifer, clogging will start halfway the well screen, and expand upward. If the interface is near the bottom of the aquifer, clogging will start near the bottom of the well and expand upward. Information on the penetration depth of oxygen in the aquifer in relation to the position of the well screen may be obtained from the concentration of oxygen in the abstracted groundwater, see Figure 8.3.

8.3 DEGASSING OF CARBON DIOXIDE AND OTHER GASSES FROM GROUNDWATER

8.3.1 Introduction

Clogging of wells due to precipitation of calcium carbonate ($CaCO_3$) is usually represented as:

$$Ca^{2+} + 2HCO_3^- \longrightarrow CaCO_3\downarrow + CO_2\uparrow + H_2O \tag{8.1}$$

In Eq. (8.1), calcite precipitation is initiated by a decrease in carbon dioxide concentration. Gasses may escape from groundwater either by diffusion or as gas bubbles. Diffusion of gasses across the water-atmosphere interface is slow (Stumm and Morgan 1981). Moreover, the conditions for diffusion are unfavorable as the annular space between well casing and pump shaft and the concentration gradient between waterlevel and well screen are both small. Consequently, a rapid decrease in gas concentration is only possible by formation of gas bubbles. The formation of gas bubbles is ascribed to reduction of pressure (Wicklein and Steuszloff 2006; Driscoll 1986) or to turbulent flow resulting in a pressure drop (Houben and Treskatis 2007; Detay 1997), but may also occur by heating around the submersible pump, in particular after a long pumping period.

Moreover, to result in precipitation of calcium carbonate, the decrease in the concentration of carbon dioxide should be sufficiently large to result in ample super saturation. This is only possible if the abstracted groundwater is already saturated, or nearly saturated, with respect to calcium carbonate. However, incrustation of wells by calcium carbonate precipitates is not a widespread phenomenon (Houben and Treskatis 2007), indicating that both conditions are not often fulfilled simultaneously.

8.3.2 Conditions for gas bubble formation

Under ideal conditions, gas bubbles may develop as soon as the sum of the partial pressures of all dissolved gasses in water exceeds the ambient pressure, i.e. the sum of hydrostatic and atmospheric pressure and the surface tension of the bubble:

$$P_{gas,total} > P_{watercolumn} + P_{atm} + P_{bubble} \tag{8.2}$$

with:

$P_{gas,total}$: sum of partial pressures of all gasses in water (atm)
$P_{watercolumn}$: hydrostatic pressure (depth below water table) (atm)
P_{atm}: atmospheric pressure (atm)
P_{bubble}: surface tension of the gas-water interface (atm)

The surface tension of a gas bubble equals (Douglas et al. 2001):

$$P_{bubble} = \frac{2\sigma}{r_{bubble}}$$

with:

σ: surface tension of gas-water interface ($N \cdot m^{-1}$)
r_{bubble}: radius of the gas-bubble (m)

The surface tension of water at 10°C equals $7.48 * 10^{-2}$ N/m. Thus, for a bubble with a diameter of 0.1 mm, the pressure inside the bubble must exceed 2908 N/m^2 = 0.29 m water column ≈ 0.029 atm. Compared to other values, this pressure is negligible. Disregarding the contribution of water vapor pressure (=0.012 atm at 10°C) and introducing Henry's relationship between concentration in water and pressure, $c_{x(aq)} = K_{H,x}P_x$, the sum of all partial gas pressures reads:

$$P_{gas,total} = \sum P_x = \sum \frac{c_{x(aq)}}{K_{H,x}} \tag{8.3}$$

with:

P_x: gas pressure (atm), where x stands for N_2, O_2, CO_2, H_2S or CH_4. The presence of trace gasses has been disregarded.
$c_{x(aq)}$: concentration of gas dissolved in water (mol/l)
$K_{H,x}$: Henry gas coefficient ($mol.l^{-1} \cdot atm^{-1}$)

Precipitation water recharging groundwater is involved in a number of redox reactions (Stumm and Morgan 1981; Appelo and Postma 2006), consuming oxygen completely and producing carbon dioxide, nitrogen, hydrogen sulfide and methane. Of these gasses, hydrogen sulfide is involved in various soil reactions, which remove this gas almost completely from groundwater. In addition, infiltrating water is involved in soil weathering reactions (Appelo and Postma 2006), consuming carbon dioxide.

Due to oscillations of the groundwater table, the concentration of gasses in groundwater may be somewhat higher than those corresponding with (soil) atmospheric conditions. This effect is called "excess air" (Holocher et al. 2003) and has been neglected as well.

Neglecting the contribution of oxygen and hydrogen sulfide to the sum of the partial pressures, and substituting the Henry constants for the remaining gasses at 10°C (Kavanaugh and Trussell 1980) in Eq. (8.2) yields:

$$P_{N_2} + P_{CO_2} + P_{CH_4} = \frac{c_{N_{2(aq)}}}{0.00077} + \frac{c_{CO_{2(aq)}}}{0.034} + \frac{c_{CH_{4(aq)}}}{0.0018} > P_{watercolumn} + P_{atm} \tag{8.4}$$

The contribution of oxygen may be neglected, as high gas pressures are caused by nitrogen, after denitrification, and methane, after methanogenesis, after oxygen has been consumed. Figure 8.4 shows for each of the gasses separately the relationship between concentration and pressure of the water column at saturation. In this figure, dissolved gas concentrations above the corresponding equilibrium relation cannot exist, since there will be a separation into two phases, i.e. gas and water. Concentrations below this relation are only present in the water phase.

In order to make this graph applicable for practical use, the pressure has been expressed in meters water column instead of atmospheres (1 atm = 10.13 m water column).

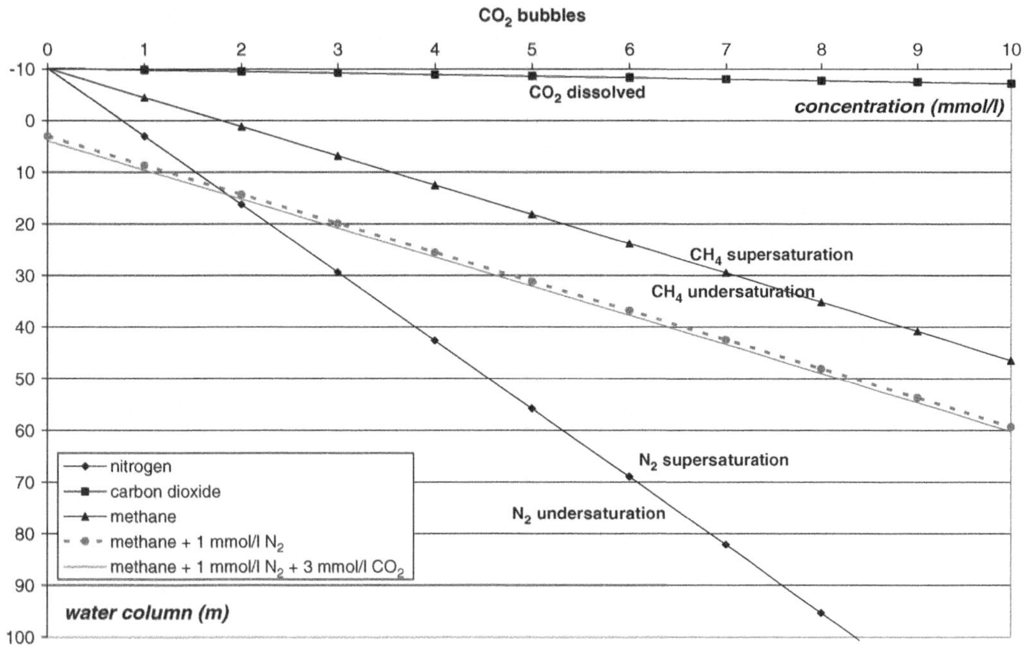

Figure 8.4 Relationship between concentration (mmol/l) of individual dissolved gasses and pressure (m water column) at saturation at 10°C. This relationship is also shown for methane together with 1 mmol/l of nitrogen and for methane together with 1 mmol/l of nitrogen and 3 mmol/l of carbon dioxide

From Eq. (8.4) it can be derived that under atmospheric conditions ($P_{N2} = 0.78$, $P_{CH4} = 0$, $P_{\text{water column}} = 0$, $P_{atm} = 1$), groundwater will be saturated or supersaturated with respect to carbon dioxide if the concentration exceeds $0.034 * (1–0.78) = 7.48$ mmol/l.

To place this value into context, due to respiration and mineralization of organic matter, the partial pressure of carbon dioxide in the root zone may reach values of more than 0.01 atm (Appelo and Postma 2006). A partial pressure of carbon dioxide of 0.01 atm is equivalent to a concentration of $c_{CO2(aq)} = K_{H,CO2}\ P_{CO2} = 0.034 * 0.01 = 0.34$ mmol/l, while the atmospheric partial pressure is equivalent to $0.034 * 0.00035 = 0.012$ mmol/l (≈0.5 mg/l). Moreover, if all oxygen introduced in groundwater were converted into an equimolar amount of carbon dioxide, this would, after substitution of the Henry constant at 10°C for oxygen (Kavanaugh and Trussell 1980), result in a concentration of $c_{CO2(aq)} = c_{O2(aq)} = K_{H,O2}\ P_{O2} = 0.0016 * 0.21 = 0.34$ mmol/l. Consequently, as a result of these processes the concentration of carbon dioxide usually will not exceed 0.5 to 1 mmol/l, and groundwater will remain under saturated with respect to carbon dioxide.

Very high concentrations of carbon dioxide in groundwater may be present in regions with post-volcanic activity, like Lake Nyos (Evans et al. 1993) or Mammoth Mountain (Evans et al. 2002). Under these conditions ($c_{N2(aq)} = 0$, $c_{CH4(aq)} = 0$, $P_{\text{water column}} = 0$, $P_{atm} = 1$), according to Eq. (8.4), groundwater will be supersaturated with respect to carbon dioxide if the concentration exceeds $0.034 * 1 = 34$ mmol/l.

Figure 8.4 shows the relationship between nitrogen concentration and pressure at saturation, where P_{CO2} and P_{CH4} have been disregarded. From this figure it is clear that nitrogen is susceptible to bubble formation: disregarding the presence of other dissolved gasses, at a depth of circa 3 m below the water table, a maximum of 1 mmol/l (28 mg/l) nitrogen can be in solution, and at about 16 m, a maximum of

2 mmol/l. At higher concentrations, nitrogen gas bubbles will develop. Of this concentration, $c_{N2(aq)} = K_{H,N2} P_{N2} = 0.00077 * 0.78 = 0.60$ mmol/l is already supplied by atmospheric nitrogen, the remainder must be furnished by denitrification.

Figure 8.4 also shows that methane is less susceptible to bubble formation than nitrogen, but is much more susceptible than carbon dioxide. This can also be deduced from the value of the Henry constants: at equal pressures, methane is about 2 times more soluble in water than nitrogen, and carbon dioxide about 20 times more soluble than methane. Disregarding the presence of other dissolved gasses, Figure 8.4 shows that groundwater at 12.5 m below the water table is able to contain a maximum of 4 mmol/l methane. Higher concentrations at that depth result in separation of phases: water with dissolved methane and gas bubbles containing methane. Figure 8.4 also shows the pressure at saturation for methane together with 1 mmol/l nitrogen, and with 1 mmol/l nitrogen plus 3 mmol/l carbon dioxide. This figure again confirms that the contribution of carbon dioxide to bubble formation is negligible. To produce a decrease in concentration as a result of bubble formation, carbon dioxide has to join other gasses.

In the Netherlands abstracted groundwater may contain up to 1.25 mmol/l (35 mg/l) nitrogen (Speksnijder et al. 1989), up to 3 mmol/l (50 mg/l) methane (Stuyfzand et al. 1994) and up to 3 mmol/l (132 mg/l) carbon dioxide. Inserting these concentrations from different locations in Eq. (8.4) yields the total gas pressure as sum of the partial pressures:

$$P_{N_2} + P_{CO_2} + P_{CH_4} = \frac{0.00125}{0.00077} + \frac{0.003}{0.034} + \frac{0.003}{0.0018} = 1.623 + 0.088 + 1.667 = 3.38\ atm$$

This equation again shows that the contribution of carbon dioxide to gas bubble formation is negligible. Considering these ranges in the concentrations of dissolved gasses, the formation of gas bubbles in groundwater abstraction will be limited to shallow wells. This finding agrees with the data of Fortuin and Willemsen (2005), who observed the presence of nitrogen and methane gas bubbles up to a depth of 20 m below the water table. They ascribe this shallow occurrence to the presence of degradable organic material at shallow depth and the absence of this material at greater depth.

8.3.3 Discussion

In the Netherlands drawdown usually varies between 2 and 5 m. This means that gas bubbles may develop in the gravel pack and screen slots if the concentration of gasses (or a gas mixture) is just below or at saturation and during abstraction is shifted to super saturation. However, the largest pressure drop occurs during upward flow in the well screen and well casing to the (submersible) pump intake. Under these circumstances, super saturation may result in the formation of deposits inside the well screen, well casing and further downstream, but not to well clogging. Abstraction by suction results in an even greater pressure drop, and consequently in a greater chance of super saturation. In case of very large draw downs (up to 100 m), the corresponding pressure release may be such that gas bubbles develop in the aquifer and gravel pack.

In the same reasoning, gas bubble formation is favored in seepage areas, where pressure is decreasing continuously during upward flow. Consequently, groundwater abstraction in seepage areas may be more vulnerable to bubble formation than outside these areas.

According to Eq. (8.4), evasion of gasses from groundwater will be maximal when the water pressure equals zero ($P_{watercolumn} = 0 =$ water surface). This condition is fulfilled in limestone caverns and by large draw downs, up to 100 m, in fully screened wells where water flows down along the screen, or in uncased boreholes where water flows down along the well bore. This reasoning is in line with the observation that clogging by calcium carbonate is confined to wells abstracting groundwater from rock aquifers (Paul 1996; Houben and Treskatis 2007).

Clogging iron hydroxide precipitates may contain some calcium carbonate (Houben 2003), but this presence is considered to be the result of secondary, reducing, reactions which produce alkalinity. If the abstracted groundwater is near calcite saturation, this alkalinity production may result in super saturation and in precipitation of calcite.

In conclusion, turbulent flow generally has nothing to do with degassing and subsequent calcite precipitation. Consequently, avoidance of turbulent flow offers no design benefits in this regard for most wells.

8.4 LAMINAR OR TURBULENT FLOW

8.4.1 Aquifer and gravel pack

To keep things as simple as possible, the radial flux (Darcy flow velocity) towards a well may be represented as:

$$q_r = \frac{Q}{2\pi L}\frac{1}{r} \tag{8.5}$$

with:

q_r: flux (Darcy flow velocity) at distance r ($m^3\ m^{-2}\ s^{-1}$)
Q: discharge rate ($m^3\ s^{-1}$)
L: length of the well screen (or height of the aquifer) (m)
r: radius, distance from the center of the well (m)

The transition between laminar and turbulent flow is indicated by the Reynolds number, and is for porous media equal to (Todd 1980; Bear 1979):

$$\mathrm{Re} = \frac{\rho_w q_r d}{\mu} \tag{8.6}$$

with:

Re: Reynolds number (-)
ρ_w: density of water (999.7 kg m^{-3} at 10°C)
d: characteristic diameter (m)
μ: dynamic viscosity (1.308 10^{-3} kg $m^{-1}\ s^{-1}$ at 10°C)

Except for the characteristic diameter d, all parameters on the right-hand side of Eq. (8.6) are well defined.

Todd (1980) substituted for the characteristic diameter d the effective grain size d_{10}, i.e. the 10% finer than diameter. According to Todd (1980), "Darcy's law is valid for $\mathrm{Re} < 1$ and does not depart seriously up to $\mathrm{Re} = 10$. A range of values rather than a unique limit must be stated because as inertial forces increase, turbulence occurs gradually". Vukovic and Soro (1992) substituted d_{17} instead of d_{10} and consequently arrived at higher Re numbers: flow is laminar if $\mathrm{Re} < 6$ and turbulent if $\mathrm{Re} > 150$.

For the characteristic diameter, Bear (1979) substituted the mean grain size, and stated that, "… practically all evidence indicates that Darcy's law is valid as long as the Reynolds number does not exceed some value between 1 and 10". Recently, Bear (2006) stated that, "Darcy's law is valid as long as Reynolds does not exceed a value of about 1, but sometimes as high as 10".

According to Freeze and Cherry (1979), linear laminar flow is present as long as $Re < (1\text{–}10)$, becoming non-linear laminar at larger values, and being turbulent at $Re > 100$. According to Driscoll (1986), flow becomes turbulent at $Re \approx 10$. Williams (1985; see also Roscoe Moss 1990) presented a somewhat different window: below $Re \approx 15$ flow is laminar and increasing with increasing grain size above $Re \approx 45$ turbulent.

Moreover, the value of the Reynolds number for the transition from laminar to turbulent flow is not only defined by the characteristic diameter of the aquifer matrix, but also by the porosity, shape and roughness of the grains. Compared to spherical grains, the Reynolds number for the transition from laminar to turbulent flow is lower for angular grains (Williams 1981).

Summarizing: the transition of laminar to turbulent flow in aquifers is not defined by a unique Reynolds number, as this number is influenced by the definition of characteristic diameter and by the porosity, shape and roughness of the grains; if $Re < 1$ to 10 groundwater flow is completely laminar; if $Re > 100$ to 150 flow is completely turbulent; and between $Re = 1\text{–}10$ and $Re = 100\text{–}150$ there is a gradual transition from laminar to turbulent flow.

Combining Eq. (8.5) and Eq. (8.6) yields the Reynolds number as function of distance from the center of the well for various combinations of discharge rate per unit length of well screen and grain diameter of aquifer or gravel pack:

$$Re = \frac{Q}{L} d_{grain} \frac{\rho_w}{\mu} \frac{1}{2\pi r} \tag{8.7}$$

This relationship is presented in Figure 8.5.

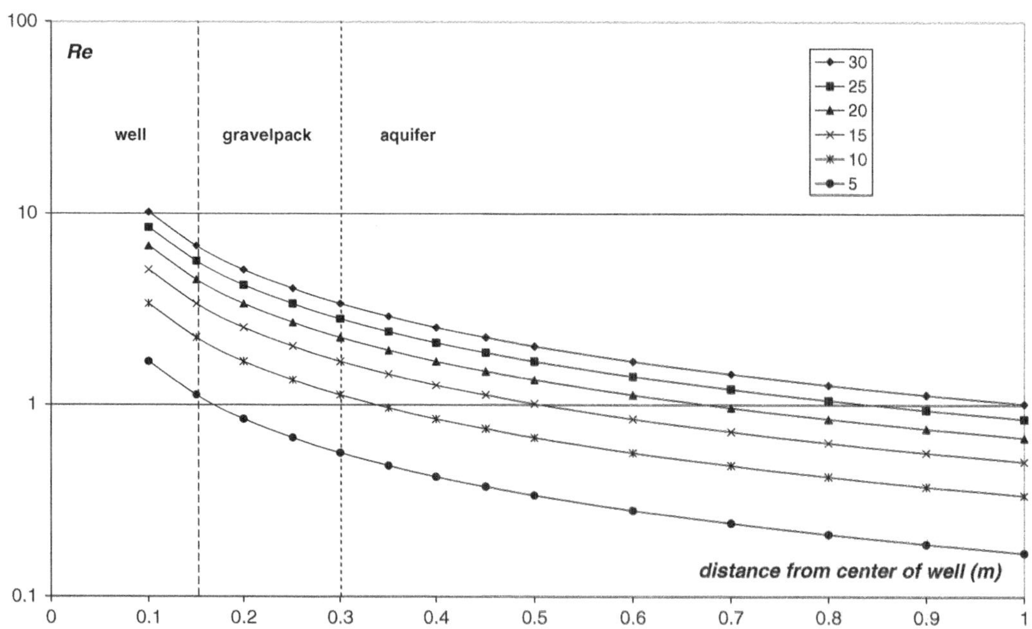

Figure 8.5 Reynolds number as function of distance from the center of the well for various combinations of discharge rate per unit length of well screen (in $m^3 \cdot h^{-1} \cdot m^{-1}$) and grain size of gravel pack or aquifer (in mm). In this figure the position of the well screen ($d_{well\ screen} = 300$ mm) and of the gravel pack/aquifer interface ($d_{well\ bore} = 600$ mm) are indicated. For convenience, $0.01\ m^3s^{-1}m^{-1} = 36\ m^3h^{-1}m^{-1}$

Figure 8.5 shows that for aquifers with a grain size of 0.5 mm and a groundwater discharge per m screen length of up to 5 $m^3h^{-1}m^{-1}$, groundwater flow on the well bore (bore diameter 600 mm) is completely laminar (Re < 1). Only in coarser aquifers (grain size > 1 mm) there would be some initial turbulence at similar discharge.

In the Netherlands, the grain size of the gravel pack varies between 1 to 3 mm, which means that under usual conditions of discharge, up to 5 $m^3h^{-1}m^{-1}$, groundwater flow in the gravel pack may show some turbulence. At similar discharge rates, the presence of a coarser gravel pack will result in more turbulence, but still not in completely turbulent flow.

In practice, however, the inflow of water over the well screen is not homogeneous. The flow velocity at both ends of the well screen may be more than twice as high as that in the middle of the screen (Nahrgang 1954; Houben 2006; Tholen 2006). Even under these conditions, fully turbulent flow is not expected.

8.4.2 Well screen slots

Next, the incoming groundwater passes the slots of the well screen. The flow velocity through the slots of the screen (=entrance velocity) equals:

$$v_{slot} = \frac{Q}{2\pi r_{screen} L} \frac{1}{\varepsilon} \tag{8.8}$$

with:

v_{slot}: flow velocity through the slots of a screen with radius r (m/s).
r_{screen}: radius of the well screen (m).
ε: fraction open area of the well screen (−).

In former times, blocking of the slots by grains of the gravel pack was quite a problem. At the moment, by application of narrower slots, and probably also by better sorting techniques, this phenomenon is minimized, and has been disregarded here.

For flow through pipes, the Reynolds number relationship reads (Douglas et al. 2001):

$$\mathrm{Re} = \frac{\rho_w v d_h}{\mu} \tag{8.9}$$

with:

v: (average) velocity (m/s).
d_h: hydraulic diameter (m).

The hydraulic diameter, d_h, of a screen slot (or pipe) equals:

$$d_h = \frac{4A}{p}$$

with:

A: area (m^2)
p: wet perimeter (m)

For a round pipe, the hydraulic diameter equals the diameter d; for a screen slot the hydraulic diameter equals:

$$d_{h,screenslot} = \frac{4w_{slot}l_{slot}}{2(w_{slot} + l_{slot})} \approx 2w_{slot} \tag{8.10}$$

with:

$d_{h,\text{ screen slot}}$: hydraulic diameter of the screen slot (m)
w_{slot}: width of the slot (m)
l_{slot}: length of the slot (m)

Substitution of Eq. (8.8) and (8.10) into Eq. (8.9) yields, after rearrangement, the Reynolds number as function of well screen diameter, discharge rate per unit length well screen, open area and slot width:

$$\text{Re} = \frac{\rho_w}{\mu} v_{slot} 2w_{slot} = \frac{Q}{L}\frac{\rho_w}{\mu}\frac{1}{\pi\varepsilon}\frac{2w_{slot}}{d_{wellscreen}} \tag{8.11}$$

with:

$d_{well\ screen}$: diameter of the well screen (m).

This relationship is presented in Figure 8.6. In circular pipes, the flow changes from laminar into turbulent at Reynolds numbers around 2000 (Douglas 2001) and becomes fully turbulent at numbers above 4000. From Figure 8.6 it is clear that for all reasonable values for discharge rate per unit length well screen (Q/L), open screen area (ε) and slot width (w_{slot}), flow through the screen slots is always laminar. Entrance velocities (=flow velocities through the screen slots) are also used as design criterion. Figure 8.6 shows that the applied values result in laminar flow.

8.4.3 Well screen and well casing

Finally, abstracted groundwater flows upwards in the well screen and the casing to the submersible pump. The flow velocity trough pipes equals:

$$v_{pipe} = \frac{Q}{\pi r_{pipe}^2} \tag{8.12}$$

Substitution of Eq. (8.12) in Eq. (8.9) and rearrangement yields:

$$\text{Re} = Q\frac{\rho_w}{\mu}\frac{4}{\pi d_{pipe}} \tag{8.13}$$

with:

d_{pipe}: diameter of the casing (and screen) (m)

Figure 8.6 also shows the Reynolds number as function of well screen diameter for several abstraction rates. From this figure it may be derived that, excepting very small abstraction rates and large diameters, vertical flow through well screens and well casings is usually turbulent.

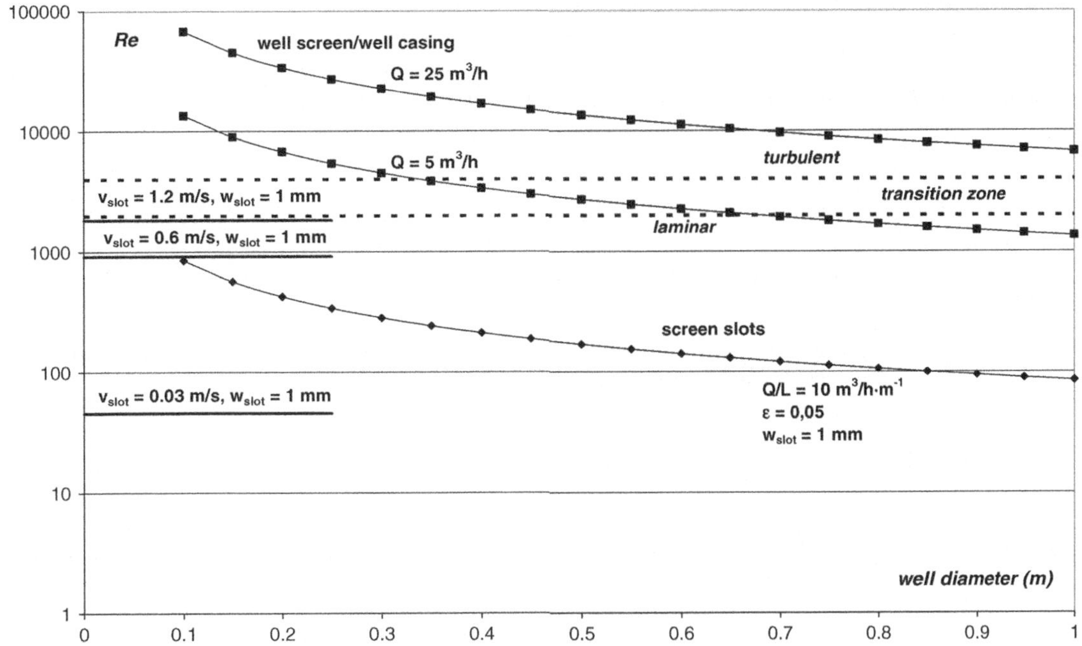

Figure 8.6 Reynolds number for horizontal flow through screen slots as function of well diameter for one set of values for discharge rate per unit length well screen (Q/L), open screen area (ε) and slot width (w_{slot}). Reynolds numbers for other values may be obtained by substitution. Also shown are the Reynolds numbers corresponding with various entrance velocities, v_{slot}. Reynolds number for vertical flow through screen and casing as function of diameter is also shown for two abstraction rates

8.4.4 Discussion

Table 8.1 summarizes the results of the calculations of the Reynolds number for a representative situation. From this table it is clear that under these conditions flow is completely laminar in the aquifer, laminar to incipient turbulent in the gravel pack, laminar in the slots, and fully turbulent during upward flow in the well screen. As the Reynolds number is linearly related to velocity (and flux), numbers for other velocities can be estimated conveniently.

8.5 WELL DESIGN VELOCITIES

8.5.1 Introduction

For the design of groundwater abstraction wells several design (maximal allowable) flow velocities are applied: through the screen slots (entrance velocity), over the aquifer-gravel pack interface (Sichardt velocity) or upward in the well.

Often a maximum entrance velocity of 0.03 m/s is advised (Driscoll 1986; Bieske et al. 1998; Ohio 2000), or 0.02 to 0.03 m/s (DVGW 2005). Screens designed to this average velocity of 0.03 m/s show an "excellent resistance to chemical encrustation" (Driscoll 1986). However, sand entrainment keeps popping up in the text. This may indicate that the authors were aware of the possible role of particles, but were unable to assess this role. At present, however, particle counters produce information about the relationship between particle concentration and groundwater flow velocity (Chapter 3 and 5). Under conditions of Table 8.1, an entrance velocity v = 0.03 m/s compares with fully laminar flow. This is confirmed by field observations and theory (Kruijtzer 1971): the difference in head between the abstraction well and the observation well in the gravel pack during operation is negligible.

Table 8.1 Summary of Reynolds numbers at various locations in and around an abstracting well. Diameter of abstraction well (d_{well}): 400 mm, bore diameter (d_{bore}): 600 mm, discharge (Q) per m screen length (L): 5 m^3/h, grain size of aquifer (d_{grain}): 500 μm, and of gravel pack (d_{gravel}): 2 mm, screen with slots 1 mm wide (w_{slot}) and 50 mm long (h_{slot}) with open area (ε): 0.05

Location	Darcy velocity or flux q ($m^3 * m^{-2} * sec^{-1}$) or pore flow velocity v (m/sec)	Reynolds number	Conclusion
Aquifer/gravel pack interface	$q_{bore} = \frac{Q}{\pi L} \frac{1}{d_{bore}} = 0.00074$	$\mathrm{Re} = \frac{\rho_w q_{bore} d_{grain}}{\mu} = 0.28$	Re < 1 à 10: flow completely laminar
Gravel pack/well screen	$q_{well} = \frac{Q}{\pi L} \frac{1}{d_{well}} = 0.0011$	$\mathrm{Re} = \frac{\rho_w q_{well} d_{gravel}}{\mu} = 1.69$	1 < Re < 10: flow almost completely laminar
Screen slot	$v_{slot} = \frac{Q}{\pi d_{screen} L} \frac{1}{\varepsilon} = 0.022$	$\mathrm{Re} = \frac{\rho_w v d_h}{\mu} \approx \frac{\rho_w v * 2 * w_{slot}}{\mu} = 33.2$	Re < 2000: flow completely laminar
Top of screen section	$v_{pipe} = \frac{Q}{\pi r^2_{pipe}} = 0.011$	$\mathrm{Re} = \frac{\rho_w v d_h}{\mu} = \frac{\rho_w v d_{well}}{\mu} = 3379$	2000 < Re < 4000: some turbulent flow

In order to maximize well efficiency, defined as the theoretical over the actual drawdown, Williams (1985; see also Roscoe Moss 1990) questioned this value of 0.03 m/s for the entrance velocity, and came up with velocities up to 0.6 to 1.2 m/s without measurable movement of sand. An entrance velocity of 0.6 m/s corresponds with laminar flow, see Figure 8.6. The entrance velocity may be recalculated to a flux (Darcy flow velocity) at distance x ($q_{r=x}$) from the heart of the well by:

$$q_{r=x} = \frac{r_{wellscreen}}{r_x} \varepsilon v_{entrance} \tag{8.14}$$

According to Eq. (8.14) an entrance velocity of 0.6 m/s together with an open screen area of 5% corresponds with a Darcy velocity (flux) at the well screen/gravel pack interface of 0.03 $m^3 m^{-2} s^{-1}$. This flux compares with Re ≈ 46, which is representative for the transition from laminar to turbulent flow. Under the conditions of Table 8.1, an entrance velocity of 0.6 m/s corresponds with a flux of 0.02 $m^3 m^{-2} s^{-1}$ at the gravel pack/well bore interface. This flux corresponds with Re ≈ 7, which is representative for incipient turbulent flow. Evidently, even the high entrance velocity of 0.6 m/s results in some turbulent flow only close to the well.

In their previous edition, AWWA (2006) advised an upper limit of 0.46 m/s for the entrance velocity, but in the current edition they acknowledge that there is no singular, uniquely defined criterion for permissible screen entrance velocity. In the selection of a design entrance velocity, aquifer characteristics and well construction method should also be considered.

After a study of several case histories of well failures due to partial clogging of well walls and screen openings, Walton (1962) concluded that the optimum screen entrance velocity should be a function of the hydraulic conductivity of the aquifer. This entrance velocity results in "avoiding migration of fine materials toward the screen and subsequent clogging of the well wall and screen openings". Figure 8.7 shows the relationship between optimum screen entrance velocity and hydraulic conductivity of the aquifer. This relationship is reasonably approached by $\sqrt{K_h}$.

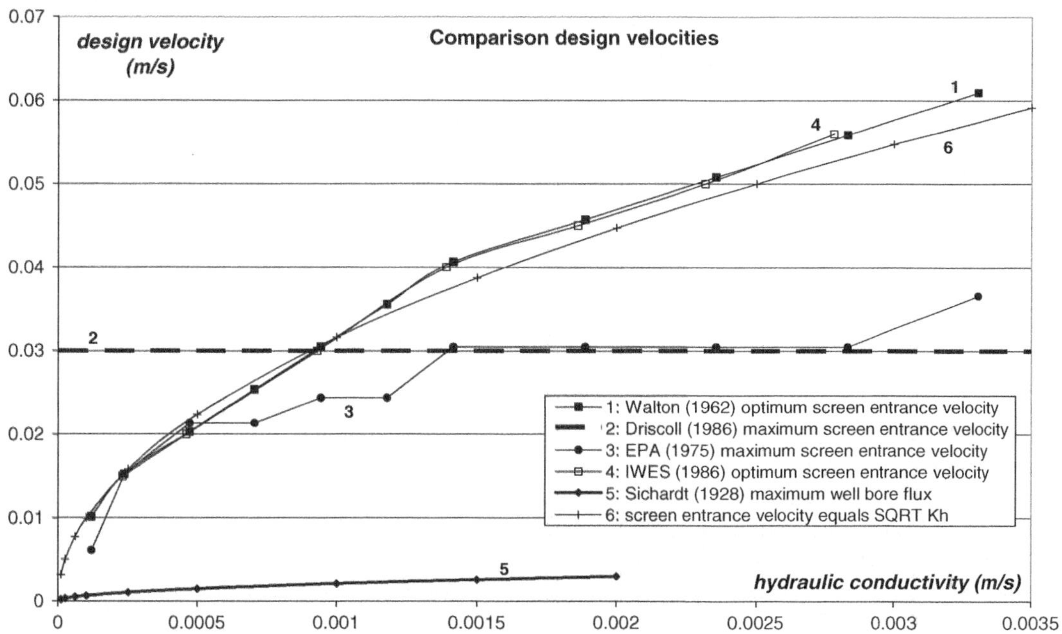

Figure 8.7 Relationship between several design entrance velocities and hydraulic conductivity

Figure 8.7 clearly shows that the optimum entrance velocity advised by the Institution of Water Engineers and Scientists (IWES) (Howsam et al. 1995) is identical to the velocity advised by Walton (1962). From this figure it is also clear that the maximum entrance velocity advised by the US Environmental Protection Agency (EPA) (Howsam et al. 1995) is a mix of both approaches: at hydraulic conductivities below circa 0.001 m/s (≈100 m/day) Walton's approach is followed and at higher hydraulic conductivities the 0.03 m/s approach.

An alternative approach for well design uses the maximum flux (Darcy flow velocity) over the gravel pack/aquifer interface (well bore). The Sichardt relation is generally applied to calculate this maximum flux (Balke et al. 2000; Bieske et al. 1998; Treskatis 1996; Olsthoorn 1976):

$$q_{max,wellbore} < \frac{\sqrt{K_h}}{15} \tag{8.15}$$

with:

K_h: hydraulic conductivity (m/s)
$q_{max,wellbore}$: maximum flux over the well bore ($m^3s^{-1}m^{-2}$)

This maximum (Sichardt) flux over the well bore is related to the maximum entrance velocity by:

$$v_{max,entrance} = \frac{1}{\varepsilon}\frac{r_{wellbore}}{r_{wellscreen}} q_{max,wellbore} < \frac{1}{\varepsilon}\frac{r_{wellbore}}{r_{wellscreen}} \frac{\sqrt{K_h}}{15}$$

Consequently Walton's entrance velocity and Sichardt's flux (Darcy velocity) over the well bore are equal if: $\frac{1}{\varepsilon}\frac{r_{wellbore}}{r_{wellscreen}} = 15$. As $\frac{r_{wellbore}}{r_{wellscreen}} \approx 1.5$, this will happen if the open area of the screen equals 10%. For smaller values, Sichardt corresponds with higher entrance velocities than Walton.

The hydraulic conductivity is related to the grain size diameter of the aquifer, for example by the Hazen relation (Williams 1981; Bieske et al. 1998): $K_h = 0.01\ d_{10}^2$, with K_h in m/s and d_{10} the 10% passing grain size in mm. Substitution of this relation in the Sichardt relation yields:

$$q_{max} < \frac{0.1 d_{10}}{15} \tag{8.16}$$

Rearranging the Reynolds relation yields:

$$q_{max} = \frac{\mu}{\rho_w d}\mathrm{Re} \tag{8.17}$$

Comparing both relations shows that in the Sichardt relation the maximum flux is linearly related with the grain size diameter, and in the Reynolds number inversely related. Consequently, the Sichardt relation does not intend to prevent turbulent flow. This is also apparent from Eq. (8.17): the larger the grain diameter, the larger the Reynolds number, and the lower the velocity where the flow changes from laminar into turbulent. See also Figure 5.12 in Roscoe Moss (1999).

Another starting point for well design is the maximal upward flow velocity of 1.22 m/s in a well as advised by AWWA (2006). Other maximum upward flow velocities mentioned are 1.5 to 2.0 m/s (Bieske et al. 1998), 1.5 m/s (Roscoe Moss 1990) and 1 m/s (DVGW 2005; Lehr et al. 1988). An upward flow velocity of 1 m/s corresponds under the conditions of Table 8.1 with an abstraction rate of circa 450 m^3/h, an unrealistic high rate.

Under the conditions of Table 8.1 and a screen length of 10 m, an upward flow velocity of 1 m/s corresponds with an entrance (screen slot) velocity of $v_{slot} = 0.2$ m/s, a Darcy velocity at the well screen/gravel pack interface of $q_{well} = 0.01\ m^3/m^2.s^{-1}$, and at the gravel pack/aquifer interface of $q_{bore} = 0.0067\ m^3/m^2.s^{-1}$. Comparison with Table 8.1 learns that the flow through the slots is laminar, and at the well screen/gravel pack (and at the gravel pack/aquifer interface) incipient turbulent.

8.5.2 Discussion

Current design rules are intended to prevent corrosion and incrustation, but there are no reports proving whether this intention is really effective. On the contrary, incrustation is occurring frequently and, with wells made of PVC, corrosion is irrelevant. This inconsistency may be ascribed to the failure to distinguish between the two types of well clogging (van Beek 1984; Chapter 7), i.e. screen slot clogging and well bore clogging. However, Walton (1962) discerns clogging of well walls and of screen openings, but does not mention different causes. Screen slot (or chemical) clogging is caused by mixing of incompatible water types and well bore (or mechanical) clogging by accumulation of particles.

Accumulation of particles on the well bore is governed 1) by the load of particles on the well bore (Chapter 3 and 5): the higher the groundwater flow velocity, the higher the particle concentration and the larger the particles; and 2) by the condition of the well bore: the smaller the diameter of the pore throats present at the well bore, the greater the chance particles cannot pass and will accumulate.

Before being taken into operation, newly built wells are developed. This development is meant to remove aquifer damage and well bore plastering present after drilling, but its success is evaluated on the absence of sand in the abstracted groundwater (DVGW (1982): < 0.01 mg/l in heavy to < 0.3 mg/l in light duty wells) or minimal turbidity (Ohio 2000). It is well known that after commissioning, new wells improve during operation. This means that the success of development should not be based on absence of sand, but on maximum specific capacity (van Beek 2002). With regard to costs, Williams (1985) arrived at the same conclusion. This is in agreement with Breedveld et al. (2007) who were able to almost double the specific capacity after usual development by the drilling company.

8.6 CONCLUSIONS

8.6.1 Development of iron hydroxide precipitates

Clogging of the screen slots by an accumulation of iron hydroxide precipitates (screen slot clogging or chemical clogging) is caused by the simultaneous abstraction of oxygen-containing and iron-containing groundwater. This condition is fulfilled by the presence of an oxygen-iron(II) redox interface in the pumped aquifer, where groundwater in the upper part of the aquifer contains oxygen and in the lower part iron(II). Precipitates may also develop in the simultaneous presence of oxygen and manganese or oxygen and methane in abstracted groundwater. During abstraction, these incompatible water qualities mix inside the well and produce precipitates above the redox interface inside the well screen and further downstream. Screen slot clogging may be characterized by chemical, hydrological, hydraulical and geological phenomena.

After switching off the pump, there is ample time for the precipitation of iron hydroxides in the mixed water containing oxygen and iron. Moreover, under phreatic conditions the filling in of the cone of abstraction results in a flow from inside the well through the slots and gravel pack in the direction of this cone. In this way, the slots and adjacent gravel pack become settled with iron hydroxides and iron oxidizing bacteria. As soon as iron hydroxide precipitates and/or iron oxidizing bacteria are present in the slots, the clogging process accelerates. This model explains the process of screen slot (or chemical) clogging conveniently: initially a lag phase and as soon as clogging get started an acceleration. This model also explains the behavior after rehabilitation: as there are inevitably some precipitates or iron oxidizing bacteria left, clogging will start immediately again without lag phase.

Continuous abstraction prevents in particular the initiation of screen slot clogging. If screen slot clogging might occur, it is limited to the inside of the well, which makes rehabilitation very effective. Consequently, mitigation of screen slot clogging has everything to do with well operation, and virtually nothing with preventing turbulent flow. This is confirmed by the widespread occurrence of well clogging due to accumulation of iron hydroxide precipitates, despite the application of current design rules. Apparently, these rules are unable to prevent well clogging.

Iron hydroxides may precipitate at some distance from the well by heterogeneous oxidation as a result of alternating oxic and anoxic conditions by filling in the cone of depression, by oxygen-iron(II) interfaces fluctuating over depth and by potential differences over the height of the aquifer during rest, and by homogeneous, heterogeneous and possibly biological oxidation during abstraction in case of a sharp oxygen-iron(II) redox interface.

8.6.2 Degassing of carbon dioxide and other gasses from groundwater

Clogging of wells by precipitation of calcium carbonate is only possible if the abstracted groundwater is at or near saturation and becomes supersaturated during abstraction. Diffusion of gasses across the water-gas interface is too slow to be of relevance for degassing in a well. Gas bubble formation is the only remaining relevant degassing process.

Carbon dioxide is too soluble in water to produce gas bubbles on its own. In order to escape in a gas bubble, carbon dioxide has to join with other less soluble gasses, specifically methane and nitrogen. The higher the concentration of these gases in water, the shallower the aquifer and the larger the drawdown, the higher the chance of gas bubble formation.

In deep wells, the pressure drop before passing the well screen is usually small compared to the drop during upward flow. This makes the likelihood of bubble formation before passing the well screen, and consequently well clogging by calcium carbonate precipitation, negligibly small. A considerable pressure relief before entering the screen is possible during prolonged upward flow to shallow wells located in seepage areas.

Exposure to atmospheric conditions (absence of a water column) is highly conducive to escape of gasses, including carbon dioxide. This condition is fulfilled by a large drawdown in fully screened wells where water flows down along the screen, or in uncased boreholes where water flows down along the well bore.

In conclusion, turbulent flow generally has nothing to do with degassing and subsequent calcite precipitation. Consequently, avoidance of turbulent flow offers no design benefits in this regard for most wells.

8.6.3 Laminar or turbulent flow

At usual abstraction rates, up to 5 m^3/h per m screen length, groundwater flow towards wells is laminar in sediments with grain size diameters up to 2 mm. Near the well screen groundwater flow may become incipient turbulent, in particular near wells with small diameters (Ø 200 to 300 mm) and coarse gravel packs ($d_{grain} > 2$ to 3 mm). Groundwater flow through the screen slots is always laminar and the upward flow in the well always turbulent. Consequently, in practice most wells meet the criterion of laminar flow towards the well.

8.6.4 Well design

Current rules applied in well design (maximum screen entrance velocity of 0.03 m/s, Walton's optimum screen velocity or Sichardt's maximum flux over the well bore and derived rules) generally result in laminar flow. Even at higher velocities, groundwater flow will still be laminar. Despite the absence of turbulent flow in the aquifer, gravel pack and screen slots, well clogging due to accumulation of iron hydroxide precipitates is a widespread phenomenon. Apparently, laminar flow is unable to prevent screen slot clogging.

Sichardt's maximum well bore flux and Walton's optimum screen entrance velocity are related to the hydraulic conductivity. Laminar flow is inversely related to the grain size diameter. Consequently, the design rules of Sichardt and Walton are not intended to maintain laminar flow: the smaller the grain size diameter, the smaller K_h and the more favorable the conditions for laminar flow.

This confusion in design rules possibly arose by not distinguishing between both types of well clogging, i.e. screen slot (or chemical) clogging and well bore (or mechanical) clogging. The occurrence of well bore clogging is governed by the load of particles on the well bore and by the filtering properties of the well bore.

The supply of particles to the well is influenced by the groundwater flow velocity. The filtering properties of the well bore are influenced by the degree of removal of aquifer damage, fine sand, wood and plant remains from drilled-through overlying layers, drilling mud and drilling additives. The more open the well bore, the easier particles may pass, especially larger particles, and the less particle accumulation occurs.

Imagine an aquifer consisting of inert glass beads of uniform diameter in its most compact structure with inert groundwater. Under these conditions no well clogging will occur. This calls for site-dependent design (see also AWWA 2006) and operation criteria:

1. well fields not vulnerable to clogging: extent of development and well operation not critical, high abstraction rates are possible.
2. well fields vulnerable to screen slot clogging: extent of development not critical, well operation critical, high abstraction rates are possible.
3. well fields vulnerable to well bore clogging: extent of development and well operation critical, abstraction rate governed by particle supply.

Information about vulnerability may be obtained from the history of the well field or a similar well field, and from chemical properties of abstracted groundwater and aquifer properties.

Chapter 9
Synthesis

C.G.E.M. van Beek
Kiwa Water Research, Groningenhaven 7, 3433 PE Nieuwegein, the Netherlands

ABSTRACT

Up till now two processes leading to clogging of groundwater abstraction wells are known: screen slot (or chemical) and well bore (or mechanical) clogging. Clogging by accumulation of calcium carbonate (calcite) is possible under very special conditions, but their occurrence is negligible.

9.1 SCREEN SLOT CLOGGING

Prevention of screen slot clogging starts already in the construction phase by preventing abstraction of two incompatible water qualities by one well. This may be done by looking for aquifers which are either completely aerobic (brown colored) or completely anaerobic (grey colored), and if only mixed aquifers are available by positioning the screen as far as possible removed from the redox interface.

Prevention, or at least mitigation, of screen slot clogging is also possible by preventing deposition of iron (III) hydroxides and attachment of Iron Oxidizing Bacteria (IOB). This may be attained by continuous abstraction resulting in a continuous inward flow. Moreover, a short residence time in the well counteracts the development of iron(III) hydroxide precipitates and adherence in the screen slots. Continuous abstraction is most conveniently achieved with the help of a variable discharge pump. A boundary condition for continuous abstraction, however, may be posed by the operation of the treatment facility of the drinking water plant or by the volume of the drinking water reservoir.

Continuous abstraction as a preventive measure was already known by experience, but this reasoning presents an explanation for this phenomenon.

9.2 PARTICLE MINERALOGY IN ABSTRACTED GROUNDWATER

Particle composition is governed by the composition of the aquifer matrix, as the source of particles, and by the chemical composition of groundwater, determining particle precipitation and particle stability. Most particles consist of common minerals like quartz and calcite, and to a lesser extent of cation alumino silicates like feldspars, muscovite, biotite, and clay minerals. Although the use of polycarbonate filters was not favorable in order to look for organic matter, there were only a few particles containing, of the elements heavier than beryllium, only carbon and oxygen.

Particles consisting of calcite were overrepresented. In order to explain this over representation it was made reasonable that these particles were continuously supplied by crushing. This phenomenon also applies to the presence of other minerals, but in particular to calcite as this is one of the minerals which is most easily crushed.

In earlier papers (van Beek and Kooper 1980; van Beek and van der Kooij 1982) well bore clogging was ascribed to the accumulation of iron sulfides and biomass of sulfate reducing bacteria (SRB), while the contribution of particles remained unclear. Afterwards it appeared that iron sulfides were a natural constituent in these aquifers (van Beek et al. 1989; Kölle et al. 1983). Sulfate reduction, producing alkalinity, may lead to calcite oversaturation and consequently to the development of calcite precipitates, comparable with the generation of ferrous iron phosphate particles (Gschwend and Reynolds 1987). However, the calcite particles encountered in this study were much larger than the iron sulfide particles, or these ferrous iron phosphate particles.

Among the most surprising observations was the presence of particles produced by corrosion and abrasion. Corrosion and abrasion particles may represent up to circa 25% of the total number of particles, and in this way influence the particle number considerably and possibly also the particle size distribution. This observation is also relevant for particle concentrations obtained from monitor wells not completely consisting of PVC, including metal spacers. Application of dedicated pumps eliminates particle contributions by the sample valve and the submersible pump, but not by the well construction.

9.3 PARTICLE CONCENTRATION IN ABSTRACTED GROUNDWATER

With measurements under minimal disturbance, Hofmann (1998) found no particles in groundwater obtained from monitor wells from a coarse aquifer on a well field, while Marre (2004) measured very high concentrations in monitor wells in the field. These high concentrations were confirmed by results of measurements with somewhat greater abstraction from monitor wells on well field Rodenhuis (water utility Oasen) and from field monitor wells by Dehnert et al. (2003). The results of these measurements cannot be biased by artifacts originating from either submersible pump or sample valve, as these measurements were executed with dedicated suction pumps. However, bias by corrosion products from well construction may have occurred. Apparently, under natural conditions, groundwater contains high particle concentrations, while around abstraction wells the particle supply seems exhausted (Hofmann 1998).

The particle concentration and particle size distribution are determined by the filtrating properties of the aquifer. The travel distance of particles in a porous medium may be described by the Yao filtration equation (Logan 1999):

$$\frac{N}{N_0} = \exp\left(-\frac{3}{2d_c}(1-\theta)\alpha\eta L\right) \tag{9.1}$$

with:

N: particle concentration (cm^{-3})
N_0: original particle concentration (cm^{-3})
d_c: grain diameter (cm)

θ: porosity
α: sticking coefficient
η: collision efficiency
L: length (distance covered) (m)

The collision efficiency is defined as:

$$\eta = \frac{\text{number of particles colliding with the collector surface}}{\text{total number of particles approaching the collector}} \tag{9.2}$$

and the sticking coefficient as:

$$\alpha = \frac{\text{number of particles depositing on the collector surface}}{\text{number of particles colliding with the collector}} \tag{9.3}$$

Actually the collision efficiency describes the physical effects on the kinetics of deposition, and the sticking coefficient the chemical effects. Consequently, in experiments with the same soil column and the same pore water velocity, the value of η remains constant. Variations in particle concentration are caused by variations in groundwater chemistry only.

Transport of particles from the bulk of flowing groundwater to the surface of the grains of the aquifer matrix occurs by diffusion, interception and sedimentation. The collision efficiency may be calculated as (Logan 1999):

$$\eta = \eta_D + \eta_I + \eta_S = 0.9\left(\frac{k_B T}{\mu d_c d_p u}\right)^{2/3} + \frac{2}{3}\left(\frac{d_p}{d_c}\right)^2 + \frac{g\Delta\rho}{18\mu u} d_p^2 \tag{9.4}$$

with:

k_B: Boltzmann constant ($1.86 * 10^{-16}$ g cm^2 s^{-2} K^{-1})
T: temperature (K)
μ: dynamic viscosity (g cm^{-1} s^{-1})
d_p: particle diameter (cm)
u: pore water velocity (cm d^{-1}) pore water velocity = q (Darcy velocity)/θ (porosity)
g: gravitation (980 cm s^{-2})
$\Delta\rho$: difference in density between particle and liquid (g cm^{-3})

Recently, Tufenkji and Elimelech (2004) presented a revised relation for the calculation of the collision efficiency version, resulting in somewhat lower values. However, the particle diameter with the minimum value for the efficiency remained the same.

Figure 9.1 shows the collision efficiency as function of particle diameter for mineral particles ($\rho_s = 2600$ kg/m^3) in a fine sandy aquifer (grain diameter 300 μm) for several pore water flow velocities. Under these conditions at low pore water flow velocities particles with a diameter of circa 0.25 μm are most mobile. With increasing flow velocity the mobility increases strongly, and the most mobile diameter shifts to slightly larger values. Increasing the grain size diameter of the aquifer matrix results in a small shift to smaller particles that are most mobile. Figure 9.1 also shows that the contribution of interception to mobility is negligible. This figure has great consequences for particle movement towards wells with steadily increasing velocities. Far away particles will hardly move, but closer to the well farther and faster after each excitation.

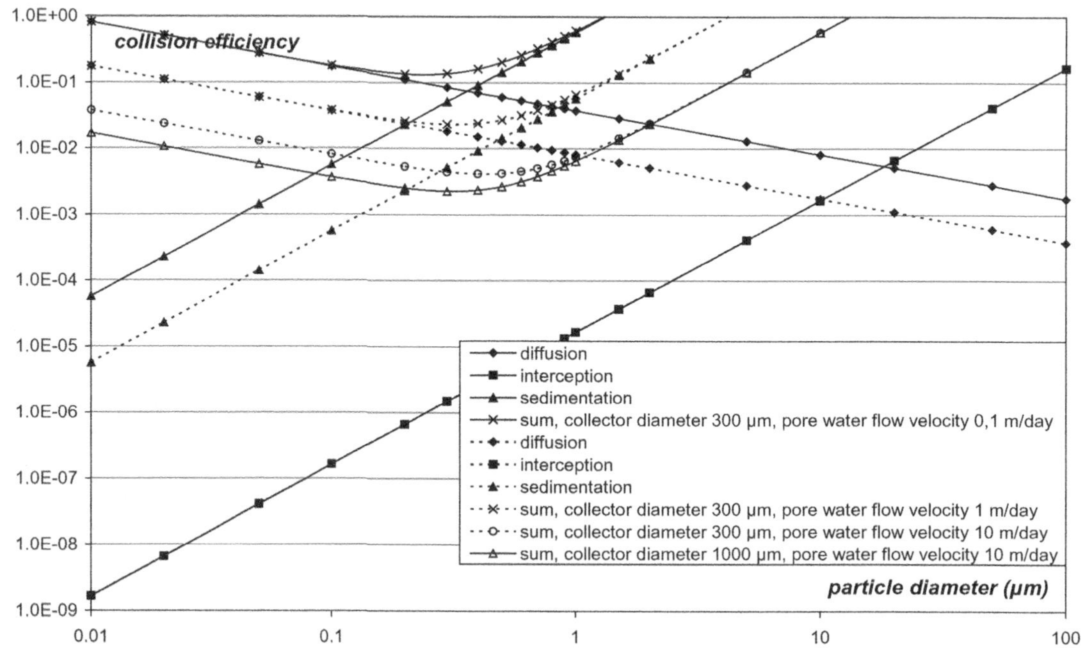

Figure 9.1 Collector efficiency as function of particle diameter for various combinations of aquifer grain (collector) diameter and pore water flow velocity for particles with a density of 2500 kg/m^3

The sticking coefficient is governed by the chemical composition of groundwater, like ionic strength and ratio of monovalent to divalent cations, and must be determined experimentally. Values in literature range from α approaching 0.01 for inert particles in groundwater with low sodium concentration to $\alpha \approx 1$ for reactive particles in groundwater with high calcium concentration (O'Melia 1990). Quartz and calcite may be considered as inert, and clay minerals as reactive particles.

Rearranging Eq. (9.1) yields the fraction of particles passing a distance L as:

$$L = -\frac{2d_c}{3(1-\theta)\alpha\eta}\ln\left(\frac{N}{N_0}\right) \tag{9.5}$$

Figure 9.2 presents the distance covered by the indicated fraction of particles as function of particle diameter for several pore water flow velocities. From this figure it is clear that the higher the pore water velocity, the greater the distance covered.

Sudden flow velocity increases resulted in instantaneous detachment of particles (Bergendahl and Grasso 2000). In these detached particles, larger particles are somewhat overrepresented (Bai and Tien 1997). Both phenomena were also encountered by sudden increases in the discharge rate during abstraction of well 50 at Noordbergum. Both phenomena will show more clearly in the closely surrounding aquifer after each time the pump starts, in particular by the acceleration of the groundwater flow, and will decrease with increasing distance to the well.

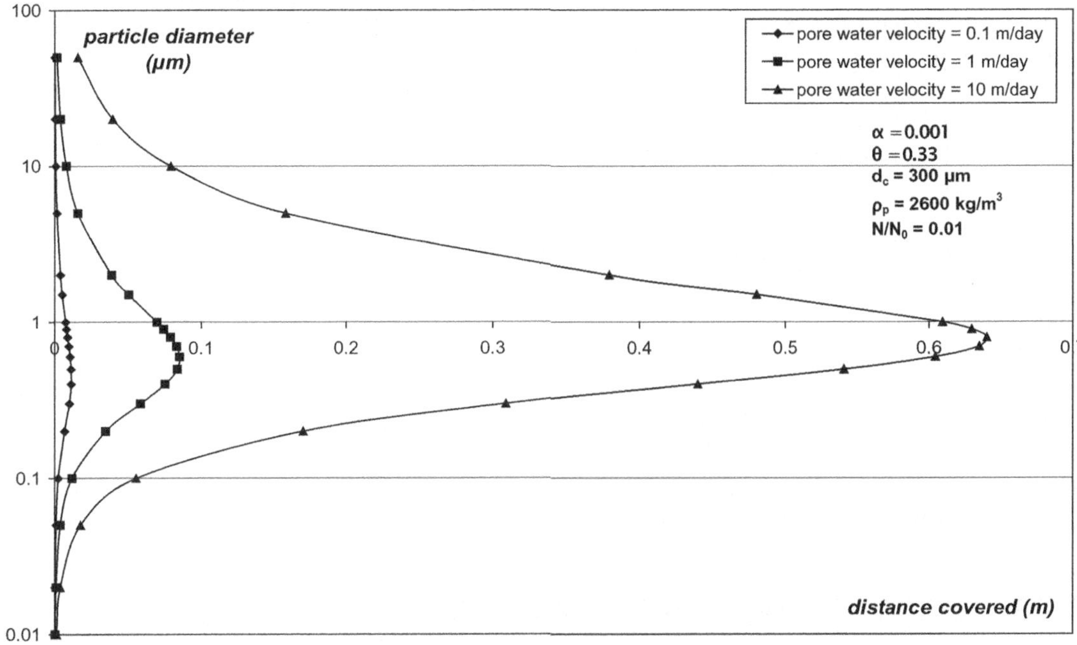

Figure 9.2 Fraction of particles (here 0.001) covering the indicated distance as function of particle diameter for several pore water velocities

As the pore water velocity increases towards the well, the closer towards the well the greater the distance detached particles will travel. Consequently, the closer to the well the more particles are removed from the particle pool than supplied, resulting in a decrease of this pool. Thus, after some time the supply of particles around the well will be exhausted, and over time this zone will move outwards.

This combination of collision and sticking may also contribute to the underrepresentation, in comparison with the Pareto relation, of particles >7 μm in abstracted groundwater. Moreover, the low sticking coefficient for inert particles may contribute to the overrepresentation of calcite and quartz particles.

In this study, particles were counted in sulfate reducing groundwater. In other studies, it was shown that aerobic groundwater contains less particles than anaerobic groundwater (Marre 2004; Ryan and Gschwend 1990). It would be interesting to obtain more information about the relation between particle concentration and redox environment.

In this study, particle concentration was counted on top of the well. However, there is no reason to believe that the particle concentration, and also the particle mineralogy, in groundwater flowing towards an abstracting well will be homogeneous in vertical and radial directions. It is recommended to measure particle concentrations over the height of the well screen with the help of a dedicated sampler. In this way the results will also not biased by possible abrasion particles from the pump and corrosion particles from the sample valve.

9.4 WELL BORE CLOGGING

The number of particles passing the well bore via pore throats, per unit time may be estimated with the help of the particle concentration in abstracted ground water and the surface area of the well bore. In sand grains in its most tight compaction, each $\frac{1}{2}d_g\sqrt{3}\ m^2$ a pore throat is present, where d_g equals the diameter of the grains. The diameter of the pore throats between these grains equals: $d_p = (2 - \sqrt{3})/(\sqrt{3})d_g = 0.155d_g$.

Assuming a (small) well bore diameter of 600 mm, a (high) abstraction rate of 5 m^3/h per m well screen, and a (high) particle concentration of 100/ml >2 μm, the number of particles passing a pore throat in a gravel pack with a grain size of 2 mm equals:

$$\frac{Q/L}{A_{wellbore}/A_{poreneck}} c_{particles} = \frac{5}{3600} \frac{\frac{1}{2} * 2 * \sqrt{3} * 10^{-6}}{2 * \pi * 0.3 * 1} 100 * 10^6 = 0.128 \frac{particles}{porethroat * s}$$

This number corresponds with one particle each 7.8 seconds. It must be realized that this load is an average value for each individual pore throat, immediately after starting the pump much higher particle concentrations are present. Moreover, the particle concentration is not homogeneously distributed vertically (de Zwart 2007) and radially over the well bore. Whether this particle load is able to result in well bore clogging depends also on the well bore characteristics. In case of a clean well bore, this load seems not insurmountable. But Timmer et al. (2003) and de Zwart (2007) showed the presence of a well damage zone, which will be much more effective in accumulating incoming particles.

9.5 PREVENTION OF WELL BORE CLOGGING

Well bore clogging may be prevented by regular intermittent abstraction. As the accumulation of particles is proportional with the abstraction time, a limit on the time span of abstraction is well conceivable. During rest, as there is no incoming flow, the particle accumulations will disintegrate. On this moment, as there is no clear picture of this disintegration process, an objective criterion for the rest time is lacking. It might be possible to shorten the rest period by reversing the flow direction. The reversed flow velocity should at least be comparable with the abstraction flow velocity. Otherwise the reversed flow will dissipate along the wide open pore throats in stead of the partly blocked ones.

Well bore clogging may also be counteracted by removing the drilling mud as complete as possible. Presently, the criterion for a developed well is the absence of sand in the abstracted water. Actually, well development should be maximal, until there is no increase anymore in specific capacity.

9.6 WELL DESIGN CRITERIA

According to literature, laminar flow would prevent screen slot clogging by incrustation. But, however flow of the incoming water generally is laminar, incrustation is a frequent phenomenon. This is not surprising, the development of chemical precipitates is determined by chemical characteristics, and not, or only to a minor degree, by hydraulic characteristics. For example, if the abstracted water does not contain both iron and oxygen, no iron hydroxides will precipitate. Laminar flow is not able to prevent this phenomenon, but, if oxygen and iron occur next to each other, turbulent flow may accelerate precipitation outside the well. Maintaining laminar upward flow inside wells is only possible at very low abstraction rates, which are not realistic.

Design criteria for wells are provided by Sichardt and Walton. Both criteria are a function of the hydraulic conductivity: the higher the hydraulic conductivity, the higher the allowable flow velocity. Higher hydraulic conductivities go together with coarser grained aquifers. Consequently, at equal Darcy flow velocities, coarse grained aquifers will show higher Reynolds numbers, and are thus more sensitive to turbulent flow. Apparently, the maintenance of laminar flow and the design criteria of Sichardt and Walton are not in line with each other.

If inert water would flow through an aquifer, consisting of inert material with uniform diameter (glass beads) in its most compact structure, nothing would happen. Apparently, Sichardt and Walton are designed to prevent transport of aquifer material. If so, both criteria should include parameters with regard to the uniformity of the grain size distribution, the density of the structure, the shape of the individual grains, etc.

References

ACE (1992). *Design, construction and maintenance of relief wells.* US Army Corps of Engineers, EM 1110-2-1914, Washington-DC.

Appelo C.A.J. and Postma D. (2005). *Geochemistry, groundwater and pollution.* Balkema, Rotterdam, 2nd ed.

Appelo C.A.J., Drijver B., Hekkenberg R. and de Jonge M. (1999). Modeling in situ iron removal from groundwater. *Groundwater*, **37** (6), 811–817.

Applin K.R. and Zhao N. (1989). The kinetics of Fe(II) oxidation and well screen encrustation. *Groundwater*, **27** (4), 168–174.

ASTM (2003). *Standard practice for characterization of particles.* F 1877–98 (reapproved 2003).

Atteia O., Perret D., Adatte T., Kozel R. and Rossi P. (1998). Characterization of natural colloids from a river and spring in a karstic basin. *Environ. Geol.*, **34**, 257–269.

AWWA (2006). *AWWA Standard Water Wells.* ANSI/AWWA A100–06.

Backhus D.A., Ryan J.N., Groher V., MacFarlane J.K. and Gschwend Ph.M. (1993). Sampling colloids and colloid-associated contaminants in ground water. *Groundwater*, **31**, 466–479.

Bai R., and Tien, C. (1997). Particle detachment in deep bed filtration. *J. Colloid Interf. Sci.*, **186**, 307–317.

Bales R.C., Li S., Yeh T.C.J., Lenczewski M.E. and Gerba C.P. (1997). Bacteriophage and microsphere transport in saturated porous media: Forced-gradient experiment at Borden, Ontario. *Water Resources Research*, **33** (4), 639–648.

Balke K.D., Beims U., Heers F.W., Hölting B., Homrighausen R. und Matthess G. (2000). *Grundwassererschliessung. Lehrbuch Hydrogeologie, Band 4.* Borntraeger, Stuttgart.

Baudisch R. (1989). Verstopfungen von Brunnenfiltern und Unterwasserpumpen durch Aluminiumoxide. *BBR*, **40** (5) 270–274.

Bear J. (2006). *Darcy's law revisited.* Lecture at Boussinecq Center for Hydrology, Amsterdam, September 1, 2006.

Bear J. (1979). *Hydraulics of groundwater.* New York, McGraw-Hill.

van Beek C.G.E.M. (2002). Mechanische putverstopping: Oorzaak, preventie en kostenbesparing. *H2O*, **35** (18) 37–39.

van Beek C.G.E.M. (1989). Rehabilitation of clogged discharge wells in the Netherlands. *Quart. J. Eng. Geol.*, **22** (1), 75–80.

van Beek C.G.E.M. (1985). Experiences with underground water treatment in the Netherlands. *Water Supply* **3 (Berlin B)**, 1–11.

van Beek C.G.E.M. (1984). Restoring well yield in the Netherlands. *J. AWWA*, **76** (10), 66–72.

van Beek C.G.E.M. (1983). *Ondergrondse ontijzering, een evaluatie van uitgevoerd onderzoek.* Kiwa Mededeling 78, Nieuwegein.

van Beek C.G.E.M. (1980). *A model for the induced removal of iron and manganese from groundwater in the aquifer.* In J. Brataschuk and J. Cole (Eds): Proc. 3rd Water-Rock Interaction Symp., Edmonton, Canada, p 29–31.

van Beek K., Oosterhof A., Breedveld R. en de Zwart B.R. (2004). Zelfregenererende pompputten, frequent schakelen voorkomt mechanische putverstopping. *H2O*, **37** (18), 36–38.

van Beek C., Schrama E., Eck M. en Verstraelen J. (1999). Verstopping van diepe putten door deeltjes. *H2O*, **32** (16/17), 21–24.

van Beek C.G.E.M., Janssen C.J.G., Juhász-Holterman M.H.A en Peters J.H. (1998). Verstopping van productieputten door deeltjes. *H2O*, **31** (17), 18–20.

van Beek C.G.E.M., Hettinga F. and Straatman R. (1989). *Release of heavy metals in groundwater due to manure spreading.* In D.L. Miles (Ed): Water rock interaction (WRI-6), Balkema, Rotterdam pp 703–706.

van Beek C.G.E.M., Hettinga F.A.M. and Straatman R. (1989). *The effects of manure spreading and acid deposition upon groundwater quality at Vierlingsbeek, the Netherlands.* In L.M. Abriola (Ed): Groundwater Contamination, IAHS Publication no 185, pp 155–162.

van Beek C.G.E.M. and van der Kooij D. (1982). Sulfate reducing bacteria in groundwater from clogging and non clogging wells in the Netherlands' river region. *Groundwater*, **20** (3), 298–302.

van Beek C.G.E.M. and Kooper W.F. (1980). The clogging of shallow discharge wells in the Netherlands river region. *Groundwater*, **18** (6), 578–586.

van Beek C.G.E.M. en Brandes M.C. (1977): Regeneratie van putten. *H2O*, **10** (24), 546–551.

Bergendahl J. and Grasso D. (2000). Prediction of colloid detachment in a model porous media: hydrodynamics. *Chem. Eng. Sci.*, **55**, 1523–1532.

Berger H., Drews M. und Lux K.N. (1995). Nachweis von Regeneriereffekten bei Brunnen der Stadtwerke Wiesbaden AG. *Bohrtechnik Brunnenbau Rohrleitungsbau*, **46** (1), 24–31.

Bieske E., Rubbert W. und Treskatis C. (1998). *Bohrbrunnen.* Oldenbourg, München, 8 Auflage.

Blackwell I.M., Howsam P. and Walker M.J. (1995a). Borehole performance in alluvial aquifers: particulate damage. *Quart. J. Eng. Geol.*, **28**, 151–162.

Blackwell I.M., Howsam P. and Walker M.J. (1995b). Permeability impairment around boreholes in micaceous aquifers. *Quart. J. Eng. Geol.*, **28**. 163–175.

Blume T., Weisbrod N. and Selker J.S. (2002). Permeability changes in layered sediments: impact of particle release. *Groundwater*, **40**, 466–474.

Böhler E. (1999). *Optimierung der Enteisenung und Entmanganung in Filteranlagen.* In: Aktuelle Themen zur Trinkwasseraufbereitung und –verteilung, Veröffentlichung aus dem Technologiezentrum Wasser Karlsruhe, DVGW-Technologiezentrum Wasser (TZW) Band 8, S 23–60.

Borch M.A., Smith S.A. and Noble L.N. (1993). *Evaluation and restoration of water supply wells.* AWWA Research Foundation, Denver, Colorado.

Borkovec M., Wu Q., Degovics G., Laggner P. and Sticher H. (1993). Surface area and size distribution of soil particles. *Colloids and Surfaces. A. Physicochemical and engineering aspects*, **73**, 65–76.

Bourgine F.P., Gennery M., Chapman J.I., Kerai H., Green J.G., Rap R.J., Ellis S. and Gaumard C. (1994). Biological processes at Saints Hill water treatment plant, Kent. *J. Institution Water Environmental Management*, **8** (4), 379–392.

Bradford S.A., Simunek J., Bettahar M., Tadassa Y.F., van Genuchten M.T. and Yates S.R. (2005). Straining of colloids at textural interfaces. *Water Resources Research*, **41**, W10404, doi:10.1029/2004WR003675.

Bradford S.A., Yates S.R., Bettahar M. and Simunek J. (2002). Physical factors affecting the transport and fate of colloids in saturated porous media. *Water Resources. Res.*, **38**, doi:10.1029/2002WR001340.

Breedveld R., van Beek K en Doedens G.J. (2007). Naar een verstoppingsvrij puttenveld Tull en 't Waal (1): Pompputten van de toekomst? *H2O*, **40** (2), 48–50.

Breeuwsma A. (1990). *Mineralogische samenstelling van de Nederlandse zand- en kleigronden.* In: W.P. Locher en H. de Bakker (Eds): Bodemkunde van Nederland, Deel 1: Algemene Bodemkunde. Malmberg, Den Bosch, the Netherlands, 2nd Ed. Ch 7, pp. 103–108.

Brodsky E.E., Roeloffs E., Woodcock D., Gall I. and Manga M. (2003). A mechanism for sustained groundwater pressure changes induced by distant earthquakes. *J. Geophysical Res.*, **108** (B8) 2390, doi:10.1029/2002JB002321.

Brown C.J., Coates J.D. and Schoonen M.A.A. (1999). Localized sulfate-reducing zones in a coastal plain aquifer. *Groundwater*, **37**, 505–516.

Brown C.J., Schoonen M.A.A. and Candela J.L. (2000). Geochemical modeling of iron, sulfur, oxygen and carbon in a coastal plain aquifer. *J. Hydr.*, **237**, 147–168.

Bruins J., Wolters H. en Wübbels G.H. (1998). Gallionella ferruginea: lust of last? *H2O*, **31** (7), 20–24.

Buchacher P., Angerer J. und Schiffer M. (2009). Sicherungsmaβnahmen nach massiven Groβhangrutsch im österreichischen Gschliefgraben. *BBR* **60** (9), 44–48.

Buffle J. and Leppard G.G. (1995), Characterization of aquatic colloids and macromolecules, 1. Structure and behavior of colloidal material. *Env. Sci. Techn.*, **29**, 2169–2175.

Bunn R.A., Magelkey R.D., Ryan J.N. and Elimelech J.N. (2002). Mobilization of natural colloids from an iron oxide-coated aquifer: effect of pH and ionic strength. *Env. Sc. Techn.*, **36**, 314–322.

Buresh R.J. and Moraghan J.T. (1976). Chemical reduction of nitrate by ferrous iron. *J. Env. Qual.*, **5**, 320–325.

Carlson L. and Schwertmann U. (1987). Iron and manganese oxides in Finnish ground water treatment plants. *Water Research*, **21**, 165–170.

Chan C.S., Fakra S.C., Edwards D.C., Emerson D. and Banfield J.F. (2009). Iron oxyhydroxide mineralization on microbial extracellular polysaccharides. *Geochim. Cosmochim. Acta*, **73**, 3807–3818, doi:10.1016/j.gca.2009.02.036.

Chanudet V. and Filella M. (2006a). A non-perturbing scheme for the mineralogical characterization and quantification of inorganic colloids in natural waters. *Env. Sc. Techn.*, **40**, 5045–5051.

Chanudet V. and Filella M. (2006b). Particle size and mineralogical composition of inorganic colloids in glacier-melting water and overlying ice in an Alpine glacier, Oberaargletscher, Switzerland. *J. Glaciol.*, **52**, 473–475.

Chanudet V. and Filella M. (2008). Size and composition of inorganic colloids in a peri-alpine, glacial flour-rich lake. *Geochim. Cosmochim. Acta*, **72**, 1466–1479.

Chapelle F. H. (1992). *Groundwater microbiology and geochemistry*. John Wiley, New York.

Cho B.Y. (2005). Iron removal using an aerated granular filter. *Process Biochem.*, **40**, 3314–3320.

Choi S., Hong S., Ahn K. and Baumann E.R. (2001). Adsorption of ferrous iron on the lepidocrocite surface. *Env. Technol.*, **22** (3), 355–365.

Close M.E., Pang L., Flintoft M.J. and Sinton L.W. (2006). Distance and flow effects on microsphere transport in a large gravel column. *J. Env. Quality*, **35** (4), 1204–1212.

Cullimore D.R. and McCann A.E. (1978). *The identification, cultivation and control of iron bacteria in ground water*. In F.A. Skinner and J.M. Shewan (Eds): Aquatic microbiology, Academic Press, New York, pp 219–261.

Czekalla Chr. (1988). *In-situ Analyse der physiologischen und filtrationstechnischen Leistung von Eisen- und Manganbakterien in Schnellfilteranlagen der Trinkwasseraufbereitung*. Dissertation Technischen Universität Braunschweig.

Czekalla Chr. (1997). Die biologische Enteisenung und Entmanganung, Verfahrenstechnik und betriebliche Aspekte. *BBR* **48** (4), 22–27.

Czekalla Chr. (2008). Mikrobiologische Grundlagen von Enteisenung und Entmanganung, ein Verfahrensüberblick. *Energie/Wasser-Praxis*, **4**, 56–58.

Czekalla Chr. und Kotulla H. (1990). Die Umstellung des Wasserwerkes Westerbeck der Stadtwerke Wolfsburg AG auf aerobe biologische Kontaktenteisenung. *GWF Wasser/Abwasser*, **131** (3), 126–132.

Czekalla Chr., Mevius W. and Hanert H. (1985). Quantitative removal of iron and manganese by microorganisms in rapid sandfilters (in situ investigations). *Water Supply*, **3** (1), 111–123.

Deelder C. (1977). Het skineffect. *H2O*, **10** (23), 522–525.

Degueldre C., Triay I., Kim J.I., Vilks P., Laaksoharju M. and Miekeley N. (2000). Groundwater colloid properties: a global approach. *Applied Geochemistry*, **15**, 1043–1051.

Dehnert J., Linkau R., Kuhn K., Neitzel P.L., Döring U. und Freyer K. (2003). Bestimmung geeigneter Förderraten bei der Grundwasserprobennahme mit hilfe von Partikel- und Radonmessungen. *Grundwasser*, **8** (2), 91–102.

Detay M. (1997). *Water wells, Implementation, maintenance and restoration*. Chicester, Wiley.

Dietz J.M. and Dempsey B.A. (2002). *Innovative treatment of alkaline mine drainage using recirculated iron oxides in a complete mix reactor*. Paper presented at the 2002 National Meeting of the American Society of Mining, Lexington, KY, June 9–13, 2002.

Dixit S. and Hering J.G. (2006). Sorption of Fe(II) and As(III) on goethite in single- and dual-sorbate systems. *Chem. Geology*, **228** (1/3), 6–15.

Diz H.R. and Novak J.T. (1998). Fluidized bed for removing iron and acidity from acid mine drainage. *J. Env. Eng.*, **124** (8), 701–708.

Douglas J.F., Gasiorek J.M. and Swaffield J.A. (2001). *Fluid mechanics*. Prentice Hall, 4th Ed.

Drever J.I. (1988). *The geochemistry of natural water*. Prentice Hall, Englewood Cliffs, NJ, 2nd ed.

Driscoll F.G. (1986). *Groundwater and wells*. Johnson Division, St. Paul, Minnesota.

Dudgeon, C.R. and Cox R.J. (1976). *Drilling mud invasion of unconsolidated aquifer materials*. Australian Government Publishing Service, Canberra.

van Duijvenbooden, W. en Breeuwsma A. (1987). *Kwetsbaarheid van het grondwater, kartering van de Nederlandse bodem in relatie tot de kwetsbaarheid van het grondwater voor verontreiniging*. Bodembescherming 65, VROM.

DVGW (1970). *Hinweise zur Verhütung der biologischen Brunnenverockerung*. DVGW Regelwerk, Merkblatt W 131, Bonn.

DVGW (1982). Arbeitsblatt W 119: *Über den Sandgehalt im Brunnenwasser; Bestimmung von Sandmengen im geförderten Wasser, Richtwerte für den Restsandgehalt*.DVGW, Bonn.

DVGW (2002). Merkblatt W 119: *Entwickeln von Brunnen durch Entsanden: Anforderungen, Verfahren, Restsandgehalte*. DVGW, Bonn.

DVGW (2005). DVGW-Regelwerk *Enteisenung und Entmanganung; Teil 1: Grundsätze und Verfahren*. Arbeitsblatt W 223–1; Bonn.

DVGW (2005). Technische Regel, *Arbeitsblatt W 118: Bemessung von Vertikalbrunnen*. Bonn.

El Azher N., Gourich B., Vial C., Soulami M.B. and Ziyad M. (2008). Study of ferrous iron oxidation in Morocco drinking water in an airlift reactor. *Chem. Engineering Processing*, **47**, 1877–1886.

Emerson D. and Weiss J.V. (2004). Bacterial iron oxidation in circumneutral freshwater habitats: Findings from the field and the laboratory. *Geomicrobiol. J.*, **21** (6), 405–414.

Enfield C.G. and Bengtson G. (1988). Macromolecular transport of hydrophobic contaminants in aqueous environments. *Groundwater*, **26**, 64–70.

Evans W.C., Kling G.W., Tuttle M.L., Tanyileke G. and White L.D. (1993). Gas buildup in Lake Nyos, Cameroon: the recharge process and its consequences. *Appl. Geochem.*, **8** (3), 207–221.

Evans W.C., Sorey M.L., Cook A.C., Kennedy B.M., Shuster D.L., Colvard E.M., White L.D. and Huebner M.A. (2002). Tracing and quantifying magmatic carbon discharge in cold groundwaters: lessons learned from Mammoth Mountain, USA. *J. Volc. Geothermal Res.*, **114**, 291–312.

Filella M., Chanudet V., Philippo S. and Quentel F. (2009). Particle size and mineralogical composition of inorganic colloids in waters draining the adit of an abandoned mine, Goesdorf, Luxembourg. *Appl. Geochem.*, **24**, 52–61.

Foppen J.W.A. (2007). *Transport of Escherichia coli in saturated porous media*. Ph.D. thesis VU University Amsterdam, the Netherlands.

Fortuin N.P.M. and Willemsen A. (2005). Exsolution of nitrogen and argon by methanogenesis in Dutch groundwater. *J. Hydr.*, **301**, 1–13.

Frankel, R.B. and Bazylinski D.A. (2003). Biologically Induced Mineralization by Bacteria. *Reviews in Mineralogy and Geochemistry*, **54** (1), 95–114.

Freeze R.A. and Cherry J.A. (1979). *Groundwater*. Prentice Hall, Englewood Cliffs, NJ.

Frischherz H., Zibuschka F., Jung H. and Zerobin W. (1985). Biological elimination of iron and manganese. *Water Supply*, **3** (1), 125–136.

Gaines R.V., Skinner H.C.W., Ford E.E., Mason B. and Rozenzweig A. (1997). *Dana's New Mineralogy*. John Wiley, New York, 8th ed.

Ghiorse W.C. (1984): Biology of iron- and manganese-depositing bacteria. Ann. Rev. Microbiol. 38, 515–550.

Goldberg S., Lebron I. and Suarez D.L. (2000). *Soil colloidal behavior.* In M.E. Sumner (Ed.): Handbook of Soil Science. CRC Press Boca Raton. Ch 6, pp. B 195–240.

Grohmann A., Gollasch R. und Schumacher G. (1989). Biologische Enteisenung und Entmanganung eines methanhaltigen Grundwassers in Speyer. *GWF Wasser/Abwasser*, **130** (9), 441–446.

Grossmann J. (2000). Regeneration von Trinkwasserbrunnen, Literaturstudie. *Gass und Wasserfach Wasser/Abwasser*, **141** (9), 586–593.

Groth P. und Czekalla Chr. (2004). *Enteisenung und Entmanganung.* in: Wasseraufbereitung - Grundlagen und Verfahren. DVGW Lehr- und Handbuch Wasserversorgung Bd. 6, Oldenbourg Industrieverlag München Wien.

Gruesbeck C. and Collins R.E. (1982). Entrainment and deposition of fine particles in porous media. *Society of Petroleum Engineers Journal*, **12**, 847–856.

Gschwend Ph.M., Backhus D.A., Macfarlane J.K. and Page A.L. (1990). Mobilization of colloids in groundwater due to infiltration of water at a coal ash disposal site. *J. Cont. Hydr.*, **6**, 307–320.

Gschwend Ph.M. and Reynolds M.D. (1987). Monodisperse ferrous phosphate colloids in an anoxic groundwater plume. *J. Cont. Hydr.*, **1**, 309–327.

Hallbeck L. and Pedersen K. (1991). Autotrophic and mixotrophic growth of Gallionella ferruginea. *Journal of General Microbiology*, **137** (11), 2657–2661.

Hanert H.H. (2002): Bacterial and Chemical Iron Oxide Deposition in a Shallow Bay on Palaea Kameni, Santorini, Greece: Microscopy, Electron Probe Microanalysis, and Photometry of in situ Experiments. *Geomicrobiology Journal*, **19**, 317–342.

Hargesheimer E.E. and Lewis C.M. (1995). *A practical guide to on-line particle counting*. American Water Works Association Research Foundation, Denver.

Harvey R.W., George L.H., Smith R.L. and Leblanc D.R. (1989). Transport of microspheres and indigenous bacteria through a sandy aquifer: results of natural- and forced-gradient tracer experiments. *Environmental Science and Technology*, **23** (1), 51–56.

Hässelbarth U. und Lüdemann D. (1967). Die biologische Verockerung von Brunnen durch Massenentwicklung von Eisen- und Manganbakterien. *BBR*, **18** (10), 363–368, (11) 401–406.

Hässelbarth U. und Lüdemann D. (1971). Die biologische Enteisenung und Entmanganung. *Vom Wasser*, **38**, 233–253.

Hermann Fr., und Tholen M. (2005): Ausschreibung von Bohr- und Brunnenbauarbeiten. *BBR*, **56** (10), 40–45.

Hiemstra T., Antelo J., Rahnemaie R. and van Riemsdijk W.H. (2010). Nanoparticles in natural systems I: The effective reactive surface area of the natural oxide fraction in field samples. *Geochimica Cosmochimica Acta*, **74**, 41–58.

Hiemstra T. and van Riemsdijk W.H. (2007). Adsorption and surface oxidation of Fe(II) on metal (hydr)oxides. *Geochim. Cosmochim. Acta*, **71**, 5913–5933.

Hiemstra T. and van Riemsdijk W.H. (2009). A surface structural model for ferrihydrite I: Sites related to primary charge, molar mass, and mass density. *Geochimica Cosmochimica Acta*, **73**, 4423–4436.

Hiemstra T. and van Riemsdijk W.H. (1996). A surface structural approach to ion adsorption: the charge distribution (CD) model. *J. Colloid Interface Sci.*, **179** (1996), 488–508.

Hiemstra T., Venema P. and van Riemsdijk W.H. (1996). Intrinsic proton affinity of reactive surface groups of metal (hydr)oxides: the bond valence principle. *J. Colloid Interface Sci.*, **184** (1996), 680–692.

Hofmann Th. (1998): *Kolloidale und suspendierte Partikel; Herkunft, Transport und Relevanz von mobilen Festphasen im Hinblick auf die künstliche Grundwasseranreicherung*. Dortmunder Beiträge zur Wasserforschung, Veröffentlichungen des Instituts für Wasserforschung GmbH Dortmund unter der Dortmunder Energie- und Wasserversorgung GmbH, Nr 56, Dortmund, Ph.D. thesis University Bremen, Germany.

Hofmann Th., Baumann Th., Bundschuh T, v.d. Kammer Fr, Leis A., Schmitt D, Schäfer Th., Thieme J, Totsche K.U. und Zänker H. (2003). Aquatische Kolloide I: Eine Übersichtarbeit zur Definition, zu Systemen und zu Relevanz. *Grundwasser*, **8** (4), 203–212.

Holocher J., Peeters F., Aeschbach-Hertig W., Kinzelbach W. and Kipfer R. (2003).: Kinetic model of gas bubble dissolution in groundwater and its implications for the dissolved gas composition. *Env. Sci Techn.*, **37** (7), 1337–1343.

Hoogendoorn J., van Paassen J. en Timmer J. (2005). Hoe een goed producerend winveld onderuit kan gaan. *H2O*, **38** (22), 35–37.

Hornberger G.M., Mills A.L. and Herman J.S. (1992). Bacterial transport in porous media: evaluation of a model using laboratory observations. *Water Resources Research*, **28** (3), 915–923.

Houben G.J. (2006). The influence of well hydraulics on the spatial distribution of well incrustations. *Groundwater*, **44** (5), 668–675.

Houben G. (2003). Iron hydroxide incrustations in wells, Part 1. Genesis, mineralogy and geochemistry. *Appl. Geochem.*, **18** (6), 927–939.

Houben G.J. and Weihe U. (2010). Spatial distribution of incrustations around a water well after 38 years of use. *Groundwater*, **48** (1), 53–58.

Houben G. and Treskatis Chr. (2007). *Water Well Rehabilitation and Reconstruction*. McGraw-Hill.

Howsam P. and Hollamby R. (1990). Drilling fluid invasion and permeability impairment in granular formations. *Quarterly J. Eng. Geol.*, **23**, 161–168.

Howsam P., Misstear B. and Jones Ch. (1995). *Monitoring, maintenance and rehabilitation of water supply boreholes*. CIRIA, London.

Huisman D.J. (1998). *Geochemical characterization of subsurface sediments in the Netherlands*. Ph.D. thesis, Wageningen Agricultural Univ., the Netherlands.

Huisman L. and Olsthoorn T.N. (1983). *Artificial groundwater recharge*. Pitman Books, Londen, UK.

Jakobson R. and Postma D. (1999). Redox zoning, rates of sulfate reduction and interactions with Fe-reduction and methanogenesis in a shallow sandy aquifer, Rømø – Denmark. *Geochim. Cosmochim. Acta*, **63**, 137–151.

James R.E. and Ferris F.G. (2004). Evidence for microbial-mediated iron oxidation at a neutrophilic groundwater spring. *Chem. Geol.*, **212** (3/4), 301–311.

Jeon B.H., Dempsey B.A. and Burgos W.D. (2003). Kinetics and mechanisms for reactions of Fe(II) with iron (III) oxides. *Env. Sci. Techn.*, **37**, 3309–3315.

Jeon B.H., Dempsey B.A., Burgos W.D. and Royer R.A. (2001). Reactions of ferrous iron with hematite. *Colloids and Surfaces A: Physicochemical Engineering Aspects*, **191**, 41–55.

Johnson D.L., Jiao J.F., DosSantos S.G. and Effler S.W. (1991). Individual particle analysis of suspended materials in Onondaga Lake, New York. *Env. Sc. Techn.*, **25**, 736–744.

de Jonge H., Jacobson O.H., de Jonge L.W. and Moldrup P. (1998). Particle-facilitated transport of prochloraz in undisturbed sandy loam soil columns. Soil Sci. Soc. Am. J., **27**, 1495–1503.

Jongmans A.G. (2004). *Waterwinput verstopping Ritskebos (Vitens), Stereomicroscopisch en micromorfologisch onderzoek*. Lab. Bodemkunde en Geologie, Wageningen Universiteit.

Jongmans A.G. (1999). *Putverstopping in het puttenveld Bergambacht van WZHH*. Lab. Bodemkunde en Geologie, Wageningen Universiteit.

Juhász-Holterman M., Balemans M. en Jansen Chr. (2004). Beter inzicht in putverstopping door flowmetingen én hoogfrequente stijghoogtemetingen. *H2O*, **37** (17), 29–31.

Kappler A. and Straub K.L. (2005). Geomicrobiological cycling of iron. *Reviews in Mineralogy and Geomicrobiology*, **59** (1), 85–108.

Katsoyiannis I.A. and Zouboulis A.I. (2004). Biological treatment of Mn(II) and Fe(II) containing groundwater: kinetic considerations and product characterization. *Water Res.*, **38** (7), 1922–1932.

Kavanaugh and Trussell (1980) in R. Sander (1999): *Compilation of Henry's Law Constants for Inorganic and Organic Species of Potential Importance in Environmental Chemistry* (Version 3) (http://www.mpch-mainz.mpg.de/~sander/res/henry.html).

Kjoeller C., Postma D. and Larsen F. (2004). Groundwater acidification and the mobilization of trace metals in a sandy aquifer. *Env. Sc. Techn.*, **38**, 2829–2835.

Kobus E.J.M. en Vlasblom W.J. (1975). *Putverstoppingen door ijzerneerslagen te Castricum*. Kiwa-Mededeling 38, Rijswijk.

Kölle W. und Schneider B. (1992). Die unterirdische Wasseraufbereitung. *Neue Deliwa Ztschr.*, (6), 272–278.

Kölle W., Strebel O. and Böttcher J. (1990). *Reduced sulphur compounds in sandy aquifers and their interactions with groundwater*. IAHS Publ. no 173, p 23–30 (Proc. Dresden Symp., March 1987).

Kölle W., Werner P., Strebel O. und Böttcher J. (1983). Denitrifikation in einem reduzierenden Grundwasserleiter. *Vom Wasser*, **61**, 125–147.

Konhauser K.O. (1997). Bacterial iron biomineralisation in nature. *FEMS Microbiology Reviews*, **20** (3/4), 315–326.

Koreman E.A. en Heuvingh W. (1998). Pelletontijzering ondersteunt milieugerichte productontwikkeling. *H2O*, **31** (12), 22–23.

Koreman E.A. en Heuvingh W. (1998). Pellet-ontijzering: milieuvriendelijker en effectief. *H2O*, **30** (4), 128.

Krems G. (1981). Brunnenalterung – Brunnenregenerierung. *Neue Deliwa Ztschr.*, **32** (7), 299–303.

Krems G. (1972). *Studie über die Brunnenalterung*. Im Auftrage des Bundesministerium des Innern, Unterabteilung Wasserwirtschaft, Berlin.

Kretzschmar R., Borkovec M., Grolimund D. and Elimelech M. (1999). Mobile subsurface colloids and their role in contaminant transport. *Adv. Agr.*, **66**, 121–194.

Kruijtzer J. (1971). Stijghoogteverliezen in en rond putfilters. *H2O*, **4** (8), 162–172.

Langmuir D. (1997). *Aqueous Environmental Geochemistry*. Prentice-Hall, Upper Saddle River, NJ.

Leake C.C. and Kamal T. (1990). *Well deterioration in Sindh, Pakistan*. In P. Howsam (Ed.): Proc. International Groundwater Engineering Conference Water wells, Monitoring, Maintenance, Rehabilitation. E.&F.N. Spon, London.

Lehr J., Hurlburt S., Gallagher B. and Voytek J. (1988). *Design and construction of water wells, A guide for engineers*. NWWA, van Nostrand Reinhold, New York.

Lerk C.F. (1965). *Enkele aspecten van de ontijzering van grondwater*. Ph.D. thesis TU Delft.

Lerman A. (1979). *Geochemical Processes, water and sediment environments*. Wiley-Interscience, New York.

Logan B.E. (1999). *Environmental Transport Processes*. John Wiley & Sons, New York.

Marre D. (2004). *Untersuchungen zum Vorkommen und Transportverhalten von Partikeln in Grundwässern und Abschätzung ihrer relevanz für den Schadstofftransport*. Ph.D. thesis Technische Universität Dresden, Deutschland.

McCarthy J.F. and Degueldre C. (1993). *Sampling and characterization of colloids and particles in ground water for studying their role in contaminant transport*. In: J. Buffle and H.P. van Leeuwen (Eds): Environmental Particles. Lewis, Boca Raton, Fla, Ch. 6 pp. 247–315.

McCarthy J.F. and McKay L.D. (2004). Colloid transport in the subsurface: past, present, and future challenges. *Vadose Zone J.*, **3**, 326–337.

McCarthy J.J. and Schamber F.H. (1981). *Least-squares fit with digital filter: a status report*. In K.F.J. Heinrich, D.E. Newbury, R.L. Myklebust and C.E. Fiori (Eds): Energy Dispersive X-ray Spectrometry. NBS Special Publication 604, Washington D.C. pp. 273–296.

McCarthy J.F. and Zachara J.M. (1989). Subsurface transport of contaminants. *Env. Sc. Techn.*, **23**, 496–502.

McDowell-Boyer L.M., Hunt J.R. and Sitar N. (1986). Particle transport through porous media. *Water Resources Research*, **22** (13), 1901–1921.

McLaughlan R. (2002). *Managing Water Well Deterioration*. IAH publication: International Contribution to Hydrogeology (ICH), volume 22. Balkema Publ, Dordrecht.

Mettler S. (2002). *In situ removal of iron from ground water: Fe(II) oxygenation, and precipitation products in a calcareous aquifer*. Ph.D. thesis Swiss Federal Institute of Technology, Zürich.

Mettler S., Abdelmoula M., Hoehn E., Schoenberger R., Weidler P. and von Gunten U. (2001). Characterization of iron and manganese precipitates from an in situ groundwater treatment plant. *Groundwater*, **39** (6), 921–930.

Mikutta Chr., Wiederhold J.G., Cirpka O.A., Hofstetter Th.B., Bourdon B. and von Gunten U. (2009). Iron isotope fractionation and atom exchange during sorption of ferrous iron to mineral surfaces. *Geochim. Cosmochim. Acta*, **73**, 1795–1812, doi:10.1016/j.gca.2009.01.014.

Mills W.B., Liu S. and Fong F.K. (1991). Literature review and model (COMET) for colloid/metals transport in porous media. *Groundwater*, **29**, 199–208.

Morgan B. and Lahav O. (2007). The effect op pH on the kinetics of spontaneous Fe(II) oxidation by O2 in aqueous solution; basic principles and a simple heuristic description. *Chemosphere*, **68**, 2080–2084.

Moser H. (1978). *Die Alterung von Vertikalbrunnen, ihre technischen und wirtschaftlichen Aspekte, aufgezeigt am Beispiel der Gewinnungsanlage des Grundwasserwerkes Mannheim-Käfertal*. Ph.D. thesis, Universität Darmstadt.

Moser H. (1979). Die Alterung von Vertikalbrunnen, ihre technischen und wirtschaftlichen Aspekte, aufgezeigt am Beispiel der Gewinnungsanlage des Grundwasserwerkes Mannheim-Käfertal. *DVGW-Schriftenreihe Wasser*, **201**, 385–418.

Mouchet P. (1992). From conventional to biological removal of iron and manganese in France. *J. AWWA*, **84** (4), 158–167.

Nahrgang G. (1954). *Zur Theorie des volkommenen und unvolkommenen Brunnen*. Springer, Berlin.

Ohio (2000). *State of Ohio, Technical Guidance for Well Construction and Ground Water Protection*. State Coordinating Committee on Ground Water.

Olsthoorn T.N. (1976). De Sichardt-snelheid en het maximaal toelaatbare debiet van waterwinningsputten. *H2O*, **9** (11), 212–214.

Olsthoorn T.N. en van Harlingen S. (1994). Infiltratieputten: schoon boren door wandschrapen. *H2O*, **27** (21), 636–639.

O'Melia C.R. (1990). *Kinetics of colloid chemical processes in aquatic systems*. In W. Stumm (Ed.): Aquatic Chemical Kinetics, Wiley, NY, p 447–474.

Pacini V.A., Ingallinella A.M. and Sanguinetti G. (2005). Removal of iron and manganese using biological roughing up flow filtration technology. *Water Research*, **39**, 4463–4475. doi:10.016/j.watres.2005.08.027.

Park U. and Dempsey B.A. (2005). Heterogeneous oxidation of Fe(II) on ferric oxide at neutral pH and a low partial pressure of O2. *Env. Sci. Technol.*, **39** (17), 6494–6500.

Paul K.F. (1996), *Brunnenalterung und Brunnenregenerierung*. In D. Flinspach und C. Castell-Exner (Eds): Wassergewinnung und Wasserwirtschaft. DVGW Lehr- und Handbuch Wasserversorgung, Bd 1, p 371–426, Oldenburg Verlag München.

Peng F., Johnson D.L. and Effler S.W. (2004). Characterization of inorganic particles in selected reservoirs and tributaries of the New York City water supply. *J. Am. Water Resources Ass.*, **40**, 663–676.

Peters J.H., Slings Q.L. en Stakelbeek A. (1992). Integrale ontwikkeling van natuur en techniek bij renovatie van een open infiltratiesysteem. *H2O*, **25** (19), 532–537.

Pettijohn F.J., Potter P.E. and Siever R. (1987). *Sand and sandstone*. Springer, New York, 2nd Ed.

Postma D. (1982). Pyrite and siderite formation in brackish and freshwater swamp sediments. *Am. J. Sc.*, **282**, 1151–1183.

Powers J.P. (1992). *Construction dewatering, new methods and applications*. John Wiley, New York.

Puls R.W. and Powell R.M. (1992). Acquisition of representative ground water quality samples for metals. *Ground Water Mon. Rev.*, **12**, 167–176.

Ouyang Y., Shinde D., Mansell R.S. and Harris W. (1996). Colloid-enhanced transport of chemicals in subsurface environments: a review. *Crit. Rev. Env. Sci. Techn.*, **26**, 189–204.

Reed M.G. (1977). Formation permeability damage by mica alteration and carbonate dissolution. *J. Petrol. Techn.*, **29**, 1056–1060.

Reddi L.N. (1997). Particle transport in soils: review of significant processes in infrastructure systems. *Journal of Infrastructure Systems*, **3** (2), 78–86.

Reimer L. (1998). *Scanning Electron Microscopy, Physics of image formation and microanalysis*. In: P.W. Hawkes (Ed.), Springer Series in Optical Sciences 45.

de Ridder D., Bruins J., Huysman K. en Kappelhof J. (2008). Biologische ontijzering aantrekkelijk voor grondwaterwinningen met twee filtratiestappen. *H2O*, **41** (16), 14–15.

Roscoe Moss Co (1990). *Handbook of Ground Water Development*. Wiley-Interscience, New York.

Roy S.B. and Dzombak D.A. (1996), Colloid release and transport processes in natural and model porous media. *Colloids and Surfaces, A. Physicochem. Eng. Aspects*, **107**, 245–262.

Ryan J.N. and Elimelech M. (1996). Colloid mobilization and transport in groundwater. *Colloids and Surfaces A: Physicochem. Eng Aspects*, **107**, 1–56.

Ryan J.N. and Gschwend Ph.M. (1990). Colloid mobilization in two Atlantic coastal plain aquifers: field studies. *Water Resources Research*, **26** (2), 307–322.

Saiers J.E. and Hornberger G.M. (1996). The role of colloidal kaolinite in the transport of cesium through laboratory sand columns. *Water Resourc. Res.*, doi.org/10.1029/95WR03096.

Saucier R.J. (1974). Considerations in gravel pack design. *Journal Petroleum Technology*, 205–212.

Schachtschabel P., Blume H.P., Brümmer G., Hartge K.H. und Schwertmann U. (1989). *Scheffer/Schachtschabel Lehrbuch der Bodenkunde*. 12. Auflage, Enke Verlag, Stuttgart.

Schamber F.H. and van Beek C.G. (2006). A 'different' kind of microscopy: Analyzing features with an automated electron beam. *Microscopy and Microanalysis*, **12** (suppl 2), 1782–1783CD.

Schijven J.F. (2001). *Virus removal from groundwater by soil passage, modeling, field and laboratory experiments*. PhD thesis Technical University Delft, the Netherlands.

Seaman J.A. (2000). Thin-foil SEM analysis of soil groundwater colloids: reducing instrument and operator bias. *Env. Sc. Techn.*, **34**, 187–191.

Seaman J.C., Bertsch P.M. and Strom R.N. (1997). Characterization of colloids mobilized from Southeastern Coastal Plain sediments. *Env. Sci. Techn.*, **31**, 2782–2790.

Sen T.K. and Khilar K.C. (2006). Review on subsurface colloids and colloid-associated contaminant transport in saturated porous media. *Advances in Colloid and Interface Science*, **119** (2/3), 71–96.

Sharma S.K. (2001). *Adsorptive iron removal from groundwater*. Ph.D. thesis, IHE Delft.

Sharma S.K., Petrusevski B. and Schippers J.C. (2005). Biological iron removal from groundwater, a review. *J. Water Supply Res. Techn. Aqua*, **54** (4), 239–247.

Sharma S.K., Kappelhof J., Groenendijk M. and Schippers J.C. (2001). Comparison of physicochemical iron removal mechanisms in filters. *J. Water Supply Res. Techn. Aqua*, **50** (4), 187–198.

Siemens J., Ilg K., Lang F. and Kaupenjohann M. (2004). Adsorption controls mobilization of colloids and leaching of dissolved phosphorus. *European J. Soil Sc.*, **55**, 253–263.

van der Sluijs P. en Locher W.P. (1990). *Korrelgrootte*. In: W.P. Locher en H. de Bakker (Eds.): Bodemkunde van Nederland, Deel 1: Algemene Bodemkunde. Malmberg, Den Bosch, the Netherlands, 2nd Ed. Ch 5, pp. 71– 90.

Smith S.A. (1992). *Methods for monitoring iron and manganese biofouling in water wells*. AWWA Research Foundation, Denver, Co.

Søgaard E.G., Aruna R., Abraham-Peskir J. and Koch Chr.B. (2001). Conditions for biological precipitation of iron by Gallionella ferruginea in a slightly polluted ground water. *Appl. Geochem.*, **16**, 1129–1137.

Søgaard E.G., Medenwaldt R. and Abraham-Peskir J.V. (2000). Conditions and rates of biotic and abiotic iron precipitation in selected Danish freshwater plants and microscopic analysis of precipitate morphology. *Water Res.*, **34** (10), 2675–2682.

Sommerfeld E.O. (1999). *Iron and manganese removal handbook*. AWWA, Denver.

Speksnijder P., Kruithof J.C., van Niekerk C.C. and van Bennekom C.A. (1989). De analyse van opgeloste gassen en de toepassing bij de grondwaterwinning en zuivering. *H2O*, **22** (4), 106–111, 114.

Stollenwerk K.G. (1996). Simulation of phosphate transport in sewage-contaminated groundwater, Cape Cod, Massachusetts. *Appl. Geochem.*, **11**, 293–298.

Straub K.L., Benz M. and Schink B. (2001). Iron metabolism in anoxic environments at near neutral pH. *FEMS Microbiology Ecology*, **34**, 181–186.

Straub K.L., Benz M., Schink B. and Widdel Fr. (1996). Anaerobic, nitrate-dependent microbial oxidation of ferrous iron. *Appl. Env. Microbiol.*, **62** (4), 1458–1460.

Stuetz R.M. and McLaughlan R.G. (2004). Impact of localized dissolved iron concentrations on the biofouling of environmental wells. *Water Sci. Technol.*, **49** (2), 107–113.

Stumm W. (1977). Chemical interaction in particle separation. *Env. Sci. Techn.*, **11** (12), 1066–1070.

Stumm W. and Morgan J.J. (1996). *Aquatic chemistry: chemical equilibria and rates in natural waters*. Wiley Interscience, New York, 3rd ed.

Stumm W. and Morgan J.J. (1981). *Aquatic Chemistry, an introduction emphasizing chemical equilibria in natural waters*. Wiley Interscience, NY, 2nd ed.

Stuyfzand P. (2007). Naar een effectievere diagnose, therapie en preventie van chemische put- en drainverstopping. *H2O*, **40** (8), 44–47.

Stuyfzand P.J. (1987). Influence of filtration and storage of groundwater samples on sample composition. *Trends Anal. Chem.*, **6**, 50–54.

Stuyfzand J.P., Lüers J. en Reijnen G.K. (1994). Geohydrochemische aspecten van methaan in grondwater in Nederland. *H2O*, **27** (17), 500–510.

Sung W. and Morgan J.J. (1980). Kinetics and products of ferrous iron oxygenation in aqueous systems. *Env. Sci. Techn.*, *14*, 561–568.

Tamura H., Kawamura S. and Hagayama M. (1980). Acceleration of the oxidation of Fe2+ ions by Fe (III)-oxyhydroxides. *Corrosion Science*, **20**, 963–971.

Tamura H., Goto K. and Nagayama M. (1976). The effect of ferric hydroxide on the oxygenation of ferrous ions in neutral solution. *Corrosion Science*, **16** (4), 197–207.

Teunissen K., Abrahamse A., Leijssen H. en van Dijk H. (2007). De wondere wereld van de deeltjes in het pompstation. *H2O*, **40** (13), 41–44.

Teutsch N., von Gunten U., Porcelli D., Cirpka O.A. and Halliday A.N. (2005). Adsorption as a cause for isotope fractionation in reduced groundwater. *Geochim. Cosmochim. Acta*, **69** (17), 4175–4185, doi:10.1016/j.gca.2005.04.007.

Tholen M. (2006). *Arbeitshilfen für den Brunnenbauer, Brunnenausbau- und Brunnenbetriebstechniken*. WVGW, Bonn.

Timmer H., Verdel J.D. and Jongmans A.G. (2003). Well clogging by particles in Dutch well fields. *J. AWWA*, **95** (8), 112–118.

Todd D.K. (1980). *Groundwater Hydrology*. John Wiley, New York, 2nd ed.

Tremblay C.V., Beaubien A., Charles Ph. and Nicell J.A. (1998). Control of biological iron removal from drinking water using oxidation-reduction potential. *Water Sci. Technol.*, **38** (6), 121–128.

Treskatis Chr. (1996). *Brunnendimensionierung und –ausbau*. In D. Flinspach und C. Castell-Exner (Eds): Wassergewinnung und Wasserwirtschaft. DVGW Lehr- und Handbuch Wasserversorgung, Bd 1, pp 199–236, Oldenburg Verlag München.

Treskatis Chr. and Leda H.D. (1998). Brunnenregenerierung - Anwendung mechanischer und chemischer Verfahren am Beispiel von Flach- und Tiefbrunnen am linken Niederrhein. *BBR*, **49** (9), 33–39.

Tüfekci N., and Sarikaya H.Z. (1996). Catalytic effect of high Fe(III) concentration on Fe(II) oxidation. *Water Sci. Techn.*, **34** (7/8), 389–396.

Tufenkji N. and Elimelech M. (2004). Correlation equation for predicting single-collector efficiency in physicochemical filtration in saturated porous media. *Env. Sci. Techn.*, **38**, 529–536.

Tyrrel S.F. and Howsam P (1997). Aspects of the occurrence and behaviour of iron bacteria in boreholes and aquifers. *Quarterly J. Eng. Geology Hydrogeology*, **30** (2), 161–169.

Tyrrel S.F. and Howsam P. (1994). Field observations of iron biofouling in water supply boreholes. *Biofouling*, **8** (11), 65–59.

Vaidya R.N. and Fogler H.S. (1990). Fines migration and formation damage; influence of pH and ion exchange. *Colloids and Surfaces*, **50**, 215–229. (In SPE Symposium on Formation Damage Control, 9th Conference, 125–132.)

de Vet W.W.J.M., Rietveld L.C. and van Loosdrecht M.C.M. (2009). Influence of iron on nitrification in full-scale drinking water filters. *J. Water Supply: Research and Technology-Aqua*, **58** (4), 247–256.

de Vet W.W.J.M., Burger W., Wolthoorn A. en van der Woerdt D. (2002). Methaanbelasting irrelevant voor filterwerking. *H2O*, **35** (1), 26–29.

Viswanathan M.N. and Boettcher B. (1991). Biological removal of iron from groundwater. *Water Sci. Techn.*, **23**, 1437–1446.

Vukovic M. and Soro A. (1992). *Hydraulics of water wells, theory and application*. Water Resources Publications, Littleton, Colorado.

Walton W.C. (1962). *Selected analytical methods for well and aquifer evaluation*. Illinois State Water Survey, Bulletin 49.

Wehrli B. (1990): *Redox reactions of metal ions at mineral surfaces*. In W. Stumm (Ed.): Aquatic chemical kinetics, p311–336. Wiley-Interscience, New York.

Wehrli B., Sulzberger B. and Stumm W. (1989). Redox processes catalyzed by hydrous oxide surfaces. *Chem. Geol.*, **78**, 167–179.

Wicklein A. und Steuβloff S. (2006). *Brunnen, ein kompexes System, Wege und Möglichkeiten eines wirtschaftlichen Brunnenbetriebes*. Expert Verlag, Band 616.

van der Wielen P.W.J.J., Senden W.J.M.K. and Medema G.J. (2008). Removal of bacteriophages MS2 and ΦX174 during transport in a sandy anoxic aquifer. *Env. Sci. Techn.*, **42** (12), 4589–4594.

Williams D.E. (1985). Modern techniques in well design. *J. AWWA*, **77** (9), 68–74.

Williams E.B. (1981). Fundamental concepts of well design. *Groundwater*, **19** (5), 527–542.

Witte K. (1981): Leistungsabfall eines Tiefbrunnens, Untersuchungen, Erkenntnisse, Massnamen. *Neue Deliwa Ztschr.*, **32** (8), 348–349.

Wolthoorn A. (2003). *Subsurface aeration of anaerobic groundwater; iron colloid formation and the nitrification process*. Ph.D. thesis Wageningen University.

van der Wurf N., van Beek K., Breedveld R. en Doedens G.J. (2007). Naar een verstoppingsvrij puttenveld Tull en 't Waal (2): Carrousel putschakelschema. *H2O*, **40** (2), 51–53.

Zonneveld J.I.S. (1947). *The Quaternary of the Peel area and surroundings*. Ph.D. thesis, Leiden Univ., the Netherlands.

de Zwart A.H. (2007): *Investigation of clogging processes in unconsolidated aquifers near water supply wells*. Ph.D. thesis, Delft Techn. Univ., the Netherlands.

IWA Publishing's authorised EU representative for General Product Safety Regulations is Diane D'Arras, 15 rue Duret, 75116 Paris, France, e-mail: safety@iwap.co.uk.

Printed and bound by CPI Group (UK) Ltd, Croydon, CR0 4YY
12/05/2026
02108891-0016